Selling the Sea, Fishing for Power

A study of conflict over marine tenure in Kei Islands, Eastern Indonesia

Asia-Pacific Environment Monograph 8

Selling the Sea, Fishing for Power

A study of conflict over marine tenure in Kei Islands, Eastern Indonesia

Dedi Supriadi Adhuri

Australian National University

E PRESS

ANU
E PRESS

Published by ANU E Press
The Australian National University
Canberra ACT 0200, Australia
Email: anuepress@anu.edu.au
This title is also available online at: http://epress.anu.edu.au/

National Library of Australia Cataloguing-in-Publication entry

Author: Adhuri, Dedi S. (Dedi Supriadi)

Title: Selling the sea, fishing for power: A study of conflict over marine tenure in
            Kei Islands, Eastern Indonesia/ Dedi Supriadi Adhuri.

ISBN: 9781922144829 (pbk.) 9781922144836 (ebook)

Series: Asia-Pacific environment monograph ; no. 8.

Notes: Includes bibliographical references.

Subjects: Fishery management--Indonesia.
          Fishery resources--Indonesia.
          Sustainable fisheries--Indonesia.
          Marine resources--Management--Indonesia.
          Fishes--Conservation--Indonesia.

Dewey Number: 333.95615091724

Cover design and layout by ANU E Press

Cover image: Photo by Dedi Adhuri.

Contents

List of Tables

List of Maps

List of Figures

List of Plates

Foreword

James J. Fox

Dedi Adhuri's *Selling the Sea, Fishing for Power* is a book of critical importance. It addresses major issues in the management of marine resources, marshals an impressive array of diverse evidence to present its argument, and then cogently sets forth a considered approach to understanding comparable forms of engagement in local marine management. It is a convincing and relevant reminder of the pertinence of politics in coastal development.

This book is an ethnographic study of several coastal communities in the Kei Islands of eastern Indonesia. Central to Dr. Adhuri's argument is an insistence that systems of local marine resource management cannot be studied on their own, in isolation from either the complex cultural and historical conditions that give impetus to community action or from the equally complex regional and national contexts within which such action is undertaken.

Dr. Adhuri's analysis probes the concept of 'community' and questions assumptions of community coherence and unity. He examines the composition of various local communities on Kei Island and shows the range of differences that exist in them and how these differences affect marine tenure and resource management. He challenges facile ideas that contend that local populations are necessarily committed to the sustainable use of their resources, that these populations possess sufficient environmental knowledge to understand the intricacies of local ecosystems, or that a shared local knowledge of these resources even exists. Recognizing all of these limitations and highlighting the politics of local and regional competition for resources, Dr. Adhuri still points to the indispensable contribution of local knowledge and community involvement in marine management. For anyone interested in marine and coastal management, whether in Indonesia or elsewhere, this book is essential reading.

There is another important dimension to this book. To provide a full understanding of marine resource management, Dr. Adhuri has written a superb ethnographic account of contemporary village life in the Kei Islands. As such, his study points to features of social life that are distinctive to the societies of eastern Indonesia. His work thus has significant comparative value. His recognition of the rich interpretability of the narratives of origin that are used as points of social reference, his assessment of local competition for precedence, and his examination of contestation – as he argues, "marine tenure is a contested practice" – all provide a dynamic view of these so-called traditional communities. A critical insight in recognizing that the rights to use marine resources held by affines – those who have married into the community – are 'nested' within ownership rights of earlier members of the community is fundamental. This 'nesting' of rights of access and use can and often is a recursive process, creating a chain of entitlement that follows the successive pattern of marriages within the community.

Despite its specific focus, this study is framed as an analysis of problems of a general kind faced by fishing communities elsewhere in Indonesia, in the Pacific, and indeed by many such communities throughout the developing world.

The research on which this study is based was carried out over years. It involved a close personal engagement with the local communities on Kei. This extended fieldwork launched Dr. Adhuri on a research career on marine and fisheries issues. Since his fieldwork on Kei, Dr. Adhuri has carried out a range of research from Aceh to Merauke both as a consultant for WorldFish Center and as the co-ordinator of the Maritime Study Group in the Research Centre for Society and Culture of the Indonesian Institute of Sciences.

The publication of this volume in the Asia Pacific Environment Monograph Series is intended to make this important work globally available. It offers great value to a wider audience.

James J. Fox
Emeritus Professor
Resource Management in Asia-Pacific Program
College of Asia & the Pacific
The Australian National University

Acknowledgements

This book is an updated revision of my dissertation that was submitted to the Department of Anthropology, Research School of Pacific and Asian Studies, Australian National University (ANU) in 2002. Without the generous support of many people, this book could not have been finished. Therefore, although it is impossible to mention everyone, I would like to express my sincere appreciation to some of those who have helped me in various stages of the preparation of this book.

First of all, I would like to convey my gratitude to Prof. James J. Fox to whom I always refer to as *Pak* Jim. He has provided me with invaluable support from the beginning, from when I first applied for a scholarship from the Australian International Development Assistance Bureau (AIDAB—now AusAID) throughout the entire period of my study and the writing of my dissertation. It was *Pak* Jim who told me that I should have my dissertation published as soon as it was submitted. Without his continuous support, I would not have finished my studies let alone this book. Andrew Walker has done tremendous work as co-supervisor when I wrote my dissertation. If I had not worked with him, turning my thesis into a book would have been much more difficult and would have taken a longer time to finish.

Second, I would like to thank those at ANU and the Resource Management in Asia-Pacific Program (RMAP) at ANU. Colin Filer has hosted several of my visits, and Mary Walta and Dana Rawls have edited my manuscript. To both Mary and Dana, I ask their forgiveness for my delays in responding to their comments and requests. Jennifer Sheehan with ANU Multimedia Services has done an excellent job creating the images and maps within the manuscript. Marshall Clark from the Institute for Professional Practice in Heritage and the Arts (IPPHA), ANU College of Arts and Social Sciences, gave me the opportunity to work on my book after I attended his workshop in February 2012.

At the WorldFish Center, I would like to express my appreciation to Eddie Allison and Maripaz Perez. When I was still working for them, they allowed me to leave Penang to work in Canberra. At the Research Center for Society and Culture (PMB) at the Indonesian Institute of Sciences (LIPI), I would like to thank Pak Hisyam, Pak Patji, and Pak Endang Turmudi for their support as the heads of PMB-LIPI.

I'd also like to acknowledge the following individuals who have supported me in several ways. Prof. Conner Bailey, Prof. Charles Zerner, and Prof. Kenneth Ruddle have allocated their time to read and provide important comments to the development of this book. In 2009, I had the opportunity to go back to the Kei Islands where Mahmud Bugis (who passed away on 9 November, 2012), my adopted family in Dullah Laut, and most of the villagers there fed me with information to catch up with the development after I left the islands in 1997. Their hospitality made me felt as though I'd gone back home.

I would also like to acknowledge the following institutions that have provided some financial support for the preparation of this book. RMAP and IPPHA at ANU financed my trip and stay in Canberra, and the WorldFish Center paid for my time while I worked in Canberra in 2010. When I was away from Jakarta, PMB-LIPI also provided financial support in the form of my salary. I would like to thank these groups for their contribution.

Last but not least, I would like to express my sincere gratitude to my wife, Tutut Sri Subekti and to my children Pratama Nurmalik Adhuri, Iesha Kinanti Adhuri, and Amyra Rachmi Adhuri for bearing with me and my many deadlines during this time.

Glossary

Dutch	English
afdelingen	department
afdelingen zuid eilanden	South Islands Residency
controleur	domestic administrator
kapitan	military commander
onderafdelingen	sub-department
posthouder	remote area representative
ratschap	kingdom
residentie	residencies
rottingknoppen	knobbed canes

Indonesian	English
adat	custom or tradition
Alivuru	pagan
anak desa	'child village'
bapak soa	father of settlement
barang bawaan	accompanying goods of the bride
bekas kebun	unused garden
belang	ceremonial war boat
bubu or fuf	box-like fish trap made from bamboo
buka	opening
bupati	head of district
camat	sub-district head
darat	land territory
desa	modern village
desa induk	centre of the village, 'mother of village'
dibuka	lifted sasi
dir-u ham-wang	precursor and carver or primus inter pares
duad	worship god
duduk adat	customary seat
fean	stake trap
gereja	church
hak makan	use right
hak milik	property right

haram	prohibited
hilaay	the great (leader)
Hukum Larvul Ngabal	Law of Red Blood and Spear from Bali
harta adat	customary wealth
iri	former slaves
kabupaten	district
kait	area of shifting cultivation
kakak	Big brother
Kalpataru	award for contribution to the environment
kaur pembangunan	development program coordinator
kaur kesejahteraan	welfare program coordinator
kaur keuangan	treasury coordinator
kaur umum	general coordinator
kebun	inland area of cultivated annual crops
kecamatan	sub-districts
kelapa	coconut
kepala desa	village head
kepala dusun	settlements head
kepala soa	traditional settlement head
kepala urusan	program coordinator
kewang	a traditional committee or individual members in charge of controlling and implementing *petuanan* and *sasi*
kintal	house and garden plot
kunci lima	five keys
labuhan	coastal area where boats are anchored
laut	sea
lebay	Islamic religious official
lembaga musyawarah desa	village deliberation council
lola	*Trochus niloticus*
lor	kingdom
lor labay	neutral kingdom
lor lim	five group
lor siw	nine group
lor utan til warat	three villages in the west
mang ohoi	wife-giving fam
mas ayam vot	gold medal in the form of crescent

mas tail til	three tail gold
mel	noble
mesjid	mosque
meti	coastal water or low tide
metu duan, tuan tan	landlord
mitu duan	attendant of the local spirit
musholla	Islamic praying place
musim	season
nangan	land
negeri	traditional village
nuhu	island
ohoi	village
ohoi kot, dusun	hamlet or settlement
ohoi murin	zone of intensive cultivation
Ohoislam	Muslim settlement
orang ilang-ilang	disappearing or invisible people
orang kaya	wealthy person, village head
panglima perang	war commander
pata lima	five group
pata siwa	nine group
pemberdayaan	empowering
perahu layar	traditional Indonesian sailing boats
perangkat desa	village functionaries
petuanan	territory under community control
petuanan kampung	settlement territory
petuanan umum	public territory
pulau kosong	uncultivated island
rahan en lim	'five houses'
ratschap	kingdom
ren	commoners, free people
riin	room of a house, segment of a fam
roa	sea territory
rok	cultivated zone crops and fruit trees
rumpun	highest traditional political organisational level
rumpun netral	neutral group
rumpun penengah	mediator group
saniri negeri	traditional village assembly

sasi	system of beliefs, rules and rituals pertaining to temporal prohibitions on use of a particular resource or territory
sekretaris desa	village secretary
sien	bad or ugly
tutup	applied sasi
lima	five
siwa	nine
ur siw	nine group
utan	kingdom
warain	forest
wilin	bridewealth
woma	centre or 'navel'
yaan	big brother
yaman	father
yan ur	wife-taking fam
yelim	donation

Abbreviations

EEZ	exclusive economic zone
HP3	*Hak Pengusahaan Perairan Pesisir* (Concessions for Marine Coastal Area')
ICG	International Crisis Group
LEK	local ecological knowledge
LMD	*Lembaga Musyawarah Desa* (Village Deliberation Council)
NGO	non-government organization
RK	*Roma Katolik* (Roman Catholic)
VOC	*Vereenigde Oost-Indische Companie* (United Dutch East Indies Company)

Currency Conversion Rates

Year	1 USD equivalent in IDR
1983	891.33
1985	1110.50
1989	1770.17
1995	2248.66
1996	2347.73
2004	8927.85

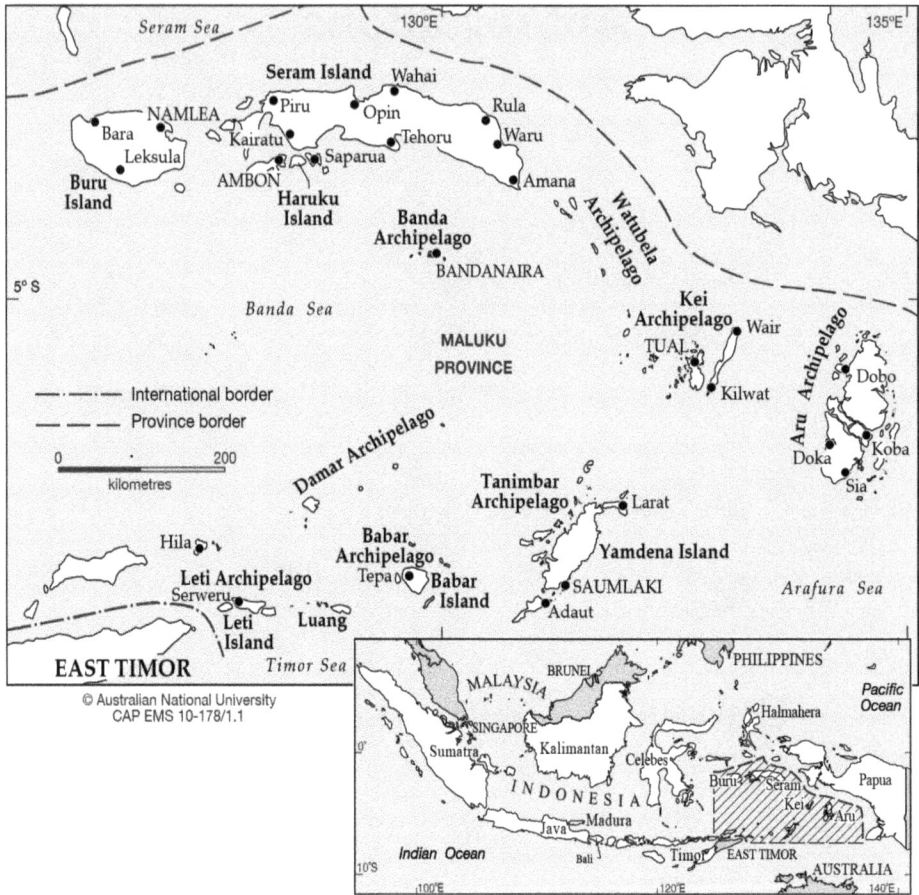

Map 1-1: The Kei Archipelago, Maluku Province, Indonesia.

Source: Education and Multimedia Services, ANU College of Asia and the Pacific.

1. Introduction

Sather, Kei Besar, April 1988. After reaching a decision the night before at a meeting in the house of the Tutrean village head, a group of 20–30 canoes—each paddled by one or two persons—headed to the coastal waters in front of Sather Village. Reaching their destination, people jumped out of their boats and started diving. It was the time to harvest Trochus. As they did a week before, the Sather villagers protested the outsiders harvesting in their territory by cursing and throwing stones at the divers. This action was motivated by their belief that the Tutrean villagers had entered and intentionally harvested the Trochus in Sather's sea territory. Being attacked by the people from Sather, the villagers of Tutrean struck back. Arrows, spears, knives and even explosives were used to fight the Sather people. Within hours many houses at Sather were burnt to the ground. The village turned red with fire and smoke covered the sky. By late afternoon, 74 houses had been reduced to ashes, another four were heavily damaged and two partially destroyed. Although there were no lives lost, almost all the wealth and property in the village houses was either destroyed by the fire or damaged by the Tutrean villagers.[1]

In the Kei Archipelago in Southeastern Maluku Province (Map 1-1), conflict over communal/traditional[2] marine tenure is commonplace. While the incident in Sather is one of the worst on record,[3] less violent conflicts have occurred in other villages (Van Hoëvell 1890; Lasomer 1985; Thorburn 2000; 2001). During 13 months of fieldwork research in 1996–97, I was witness to three such conflicts that arose in Dullah Laut Village on the Kei Kecil group of islands.

Personal experience with conflicts surrounding resource ownership and analysis of their documentation has enabled me to grasp the concept and practice of communal marine tenure and how crucial this is for understanding the human-environment relationship in the Kei Islands. As my understanding of the nature of these conflicts has developed, I have discovered that communal marine tenure is not solely a means of resource management. In the Kei Islands, communal marine tenure is a complex phenomenon that concerns the relationship between humans and their marine environment and the relationship between groups where it is used as an instrument for political positioning of entities within

---

1   This account is based on interviews and legal documents from the Maluku Tenggara Regency Court (1991).
2   The traditional marine tenure practiced in Kei Islands is communal in nature and discussed in detail in Chapter Five.
3   Another example of violent conflict occurred in 1995, between villagers from Hollat Atas and Hollat Bawah hamlets, resulting in fatalities and burning of houses (Silubun 2004: 129–30).

and between communities. As demonstrated by the case studies presented in this volume, marine tenure is either about social and political positions such as those held by the 'nobles' (*mel*) or 'free people' (*ren*), or about the authority of a village leader. In economic terms, marine tenure—as the exclusive right to gain benefit or exclude others from making use of sea territory—can not be exercised freely without intervention from institutions within and outside the community. Hence, traditional marine tenure should not be seen as a self-contained institution but rather a highly politicised mechanism for resource management.

In line with this understanding, this book proposes a different perspective on communal marine tenure from the current view that sees tenure merely as a means of marine resource management. The perspective presented here considers marine tenure in a broader social context, incorporating the ways in which traditional marine tenure is embedded in the social world of the community. Therefore, an understanding of how people perceive and practice traditional marine tenure should reflect the community social structure and in particular demonstrate the importance of 'power play' in determining marine tenure and management practice. Taking such aspects into account provides a more holistic position from which to assess how such practices will be impacted by challenges associated with modernisation and market development.

The context for the position outlined was composed from an analysis of four conflicts related to marine tenure in the Kei Islands, during the period 1996–97.[4] The first conflict demonstrated how local elites used traditional marine tenure to leverage 'political capital' in a village leadership contestation. The second conflict revealed that the legal status of customary marine tenure is not considered particularly significant, because on land or sea, the most powerful players and the institution that served their interest best, determined the rules and procedures followed. The third conflict illustrates that market economy did not necessarily degrade the practice of traditional marine tenure. In this situation, the market informed the economic value of the resources, strengthening territorial claims that provided greater access for some and excluded others. The fourth conflict shows that customary marine tenure was the subject of contestation in the context of tradition. This occurred because control over marine territory designates a precedence position. Hence, when there is a precedence contestation within traditional groups in the community, marine tenure becomes a contested issue. With respect to the function of marine tenure as a means of resource management, these conflicts clearly indicated

---

4 Cases were chosen based on the assumption that they would represent the complexity of issues relating to conflict over marine tenure in the Kei Islands. The first three conflicts relate to the practice of traditional marine tenure under contemporary political and economic situations, while the fourth case represents marine tenure within its traditional context.

the disregard people have regarding environmental and sustainability issues. This was illustrated by the lack of concern regarding cyanide fishing (in which cyanide is used to stun fish making them easier to catch) in two out of four conflict case studies. In the last conflict case study, entry to the disputed marine territory was restricted and as such, development of 'proper' marine resource management can not be verified although it was not anticipated.

The arguments presented here have been informed by the academic discourse on traditional marine resource management in Maluku and marine tenure in the context of resource management, both of which I will proceed to outline.

Traditional Marine Resource Management in Maluku

Central to traditional marine resource management in Maluku is the focus on the system of beliefs, rules and rituals pertaining to temporal prohibitions on the use of a particular resource or territory referred to as *sasi*. When *sasi* is applied (*tutup*) to a particular resource, no one, including the owner, can harvest it until the *sasi* is lifted (*dibuka*). *Sasi* applied to a coconut tree, for example, means that no one can harvest or take home fallen coconuts until the restriction is lifted. Similarly, *sasi* applied to territory restricts extraction of resources from a particular territory.

Sasi is differentiated on the basis of a specified resource or territory as well as the belief system, ritual leaders and location (see Monk et al. 1997) and Soselisa (2002) for more detailed accounts of *sasi*). Some examples of terms used for resource and territory *sasi* include those that describe: coconuts (*kelapa*); *Trochus niloticus* (*lola*); land (*darat*); and sea (*laut*). Terms related to belief systems, ritual leaders and location include: local village beliefs (*sasi negeri*); Christian rituals conducted in a church by a priest (*sasi gereja*); and Islamic rituals conducted in a mosque by an imam (*sasi mesjid*). The rituals of applying and lifting the *sasi* for local village beliefs are performed at sacred places in the village, led by a traditional leader.

In relation to traditional marine resource management, *sasi* is applied to specific areas of sea territory (*petuanan laut*), coastal waters (*meti*) or the area where people anchor their boats (*labuhan*). These areas are defined and are under the control of a particular village or social grouping. In the ritual of applying *sasi*, the ritual leader announces the sea boundaries and which sea resources in that territory are under *sasi* regulation. Resources covered by *sasi* and the gear used for extraction are also declared. In the case of local village beliefs *sasi*, violation penalties are also stipulated.

When the resources are ready for exploitation, the same ritual practitioner will perform an opening or lifting *sasi*. As well as communication with the spirit world, the ritual also informs people of how to harvest the resource, including: who can take part in the harvest; what gear can be used; what and how much of the resource can be taken; the system for distribution; and most importantly, how long the *sasi* will be opened.

The discourse on *sasi* has burgeoned since the 1980s particularly among non-government organisations (NGOs), provincial governments, and academic institutions based in Ambon. Inspired by widespread environmental and social movements, NGO workers became actively involved in empowering local leaders to revive and document management practices. A major focus of research has been identifying the elements of resource management and the distribution of such practices. These efforts were formally acknowledged when Kalpataru[5] was awarded to two villages in Maluku Tengah for practicing 'sustainable' traditional resource management. In addition, the research papers have made an important contribution to the discussion of traditional marine tenure both in Indonesia and internationally.

A report written jointly by an NGO and academic researchers from the Law Faculty and Maluku Research Centre at the University of Pattimura in Ambon, described *sasi* in the following terms:

> [*Sasi*] strongly supports conservation of living marine resources ... in addition to being rather useful because it regulates the resource use, extraction and protection; it also ensures an even distribution of the harvest. (translated from Anonymous 1991; see also UPPPSL 1995).

This description is consistent with a definition given by a traditional committee leader in charge of conducting rituals and monitoring the practice of *sasi* in Haruku Village, Central Maluku. He noted that '*sasi* can be described as a prohibition on the harvesting of certain natural resources in an effort to protect the quality and population of that biological natural resource (animal or plant)' (Kissya 1995: 4). Lokollo (1994), a legal scholar based in Ambon, also supports such definitions and he has long held the belief that *sasi* should be the basic model for the national policy on rural environmental management (Lokollo 1988).

In the early-1990s, a more critical perspective on *sasi* emerged, suggesting the above arguments might be misleading because they were constructed without reference to the historical and socio-political context of *sasi* (Pannell 1997). Taking such contexts into account, it is argued that:

---

5   Kalpataru is an award given by the Ministry of Population and Environment for major contributions to the environment.

> [s]asi has undergone considerable change over the past 400 years ...
> it has developed from a ritual protection of communal resources to a
> governmentally regulated regime of agro-ecological control of private
> and common resources, and from there to a largely commercialized
> and privatized means of theft prevention (Von Benda-Beckmann et
> al. 1992: 5).

Interestingly such historical analysis would indicate that *sasi* is a management
system largely designed by elites from inside and outside local communities. This
was illustrated during the latter stages of the colonial era with the ratification of
sasi rules initiated by traditional elites in collaboration with local Dutch officials
to meet the economic and political interests of both local and colonial elites
(Zerner 1994a: 1087). Another elite initiative altering traditional *sasi* mechanisms
occurred in relation to sea territory in Nolloth Village on Saparua Island with
the development of the Trochus market in the 1950s (Zerner 1991). The head
of Nolloth Village issued *sasi* on the village sea territory due to the increase in
demand for Trochus and at the same time instituted changes to *sasi* practices.
Previously, when restrictions were lifted, the sea territory was open to all village
community members to harvest Trochus. However, with the 'new' system of
sasi, the village headman declared the territory closed to community members
and the village administration assumed total control, allocating the income from
Trochus harvest for village programs such as roads and public toilets. Problems
quickly emerged regarding the hiring of non-villagers for harvesting Trochus
which was seen as depriving villagers of income. Villagers questioned whether
income from Trochus was intended to benefit the whole community.

Studies of contemporary practice of *sasi* provide further insights into the local
realities as noted by Pannell:

> [T]he practices referred to and associated with sasi in the marine
> environment of Luang [Southeastern Maluku] minimally involve
> the interest and actions of residents of this island, the commercial
> machinations of regional traders and internationals exporters, the
> fashions and fads of distance consumers, the compliance and blessing
> of the Church and its agents, as well as the endorsement of village
> representatives of local government institutions and the support of
> government personnel from other jurisdictions. In addition, let us not
> forget those fishermen who, though their non-sanctioned exploitation
> of local marine resources, contribute to the social delimitation of the
> efficacy of invoking sasi (Pannell 1997: 297).

Having noted the involvement and interests of various agencies, Pannell
suggests that *sasi* might mean different things to different agencies with different
interests. For example,

for the traders the opening of sasi ensures that they enjoy exclusive rights of purchase [on the harvest] ... for people on Luang, the payments made by traders [for his monopolistic rights to buy the harvest] also amount to de facto recognition of their rights and interests as customary and communal title holders of these marine areas (ibid.: 296).

Evaluating contemporary *sasi* practices in Watlaar Village on Kei Besar Island, Antunès and Dwiono found that the monopolistic control by a traditional leader had caused villagers to over-harvest the Trochus as well as question the distributional fairness of the practice (Antunès 2000; Antunès and Dwiono 1998).

These historical and contemporary analyses raise questions regarding the conservation and equity factors considered inherent to *sasi*. Analysed in its sociopolitical context, it is evident that local traditional leaders, NGOs, and scholars actively engage in the process of 'greening' *sasi*. On this point, Zerner (1994a) writes that the political context of the emergence of green *sasi* includes both a growing environmental awareness as well as resistance of local elites and NGOs to increasing resource control by the central government and fishing industry. In this sense, green *sasi* as a political discourse aims to empower marginalised local people.

In the Kei Islands, the situation differs in that people focus on communal marine tenure rather than *sasi*. The practice of *sasi* has been abandoned in most villages[6] except in some places on the east coast of Kei Besar, although even here ownership of sea territory is becoming more important. My investigations of *sasi* in Sather, Tutrean and Hollat villages in eastern Kei Besar Island (Map 1-2) were hampered due to conflict between and within these villages over sea territory. In 1979, Barraud's study of Kei Tanimbar indicated that communal marine tenure practice was not in evidence. However, in the early 1990s, conflict over the issue of sea territory broke out indicating marine tenure was much more important than *sasi*.[7]

A significant gap in the discourse on *sasi* is that there is no explanation posited for the increase in tension in relation to marine tenure. Why is this? I believe it is because the discourse on *sasi* tends to take the issue of marine ownership for granted. In the discourse on *sasi*, the issues of marine ownership—represented by concepts such as sea or coastal territory under community control—are not considered. Yet, if we look at the practice of *sasi* in Nolloth, for example, we can see that after the village head applied *sasi* on Trochus, the issue of rights over

---

6   A recent study on the 'presence, performance, and institutional resilience of *sasi*' found that it is also declining in Central Maluku (Harkes and Novaczek 2002; see also Novaczek, Harkes et al. 2001).

7   The concept of *petuanan*, which covers land and sea estates, is embedded in Kei Tanimbar tradition which would mean communal marine tenure was practiced. However, the issue of sea ownership was obviously not an important issue at the time of Barraud's study.

territory and the resource itself became contentious. The village head's declaration that the territory was under the control of the village administration—meaning the territory, or at least the harvest of Trochus, was closed to the community—was not consistent with the understanding of traditional marine tenure that existed previously. Furthermore, when the people asked if the money from the Trochus was for the whole community, they were actually questioning the right of the village head and his staff to represent the community. This suggests that understanding and practice regarding the concept of traditional marine tenure differed between the village head and members of the community and this discordance created tension and conflict.[8] The effects of different attitudes towards issues of sea ownership are evident in many of the case studies examined in this volume.

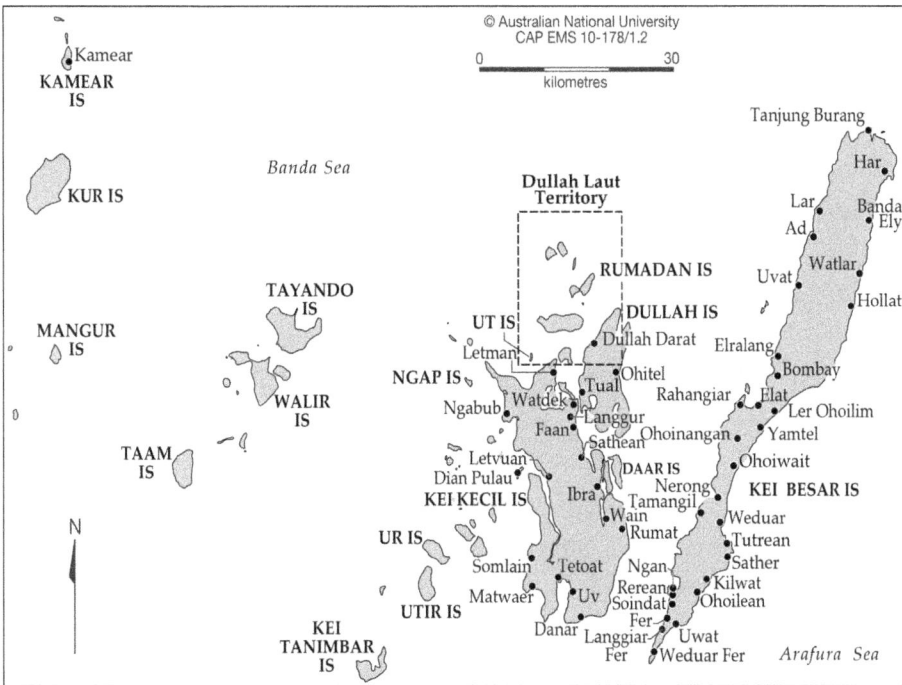

Map 1-2: The Kei Islands.

Source: Author's fieldwork.

---

8   Zerner (1996: 79) noted similar problems at Paperu and Porto villages on the same island.

Discourse on Marine Tenure and Management

The issue of ownership in relation to resource management was popularised by *The Tragedy of the Commons*. In his article, Hardin (1968) notes that resources that are not subject to ownership—what he calls common property—tend to be over-exploited. He argues that in common property situations where the resource is free for all, there is no incentive for individuals to take responsibility for the sustainability of the resource, encouraging over-exploitation and resource degradation. Hardin explains the 'tragedy' as follows:

> Adding together the component partial utilities, the rational herdsman concludes that the only sensible course for him to pursue is to add another animal to his herd. And another and another .... But that is the conclusion reached by each and every rational herdsman sharing a commons. Therein is the tragedy. Each man is locked into a system that compels him to increase his herd without limit—in a world that is limited. Ruin is the destination toward which all men rush, each pursuing his own best interest in a society that believes in the freedom of the commons. Freedom in a commons brings ruin to all (Hardin 1968: 20).

Having identified this fundamental problem, Hardin and others following the same theoretical line propose some solutions. These include creating an institution of private ownership (Gordon 1954; Demsetz 1967), sole ownership (Scott 1955),[9] and 'arrangements that create coercion ... [a] mutual coercion, mutually agreed upon by the majority of the people affected' (Hardin 1968: 26–7). The first two proposals are based on the economic calculation that whenever a resource is owned privately the owner will take into account the externalities of his exploitation. This is because the private owner will incur all economic inputs and outputs, including the cost of negative impacts. The economic considerations will motivate the owner to use his resource efficiently. Hardin's proposals are based on the idea that it is impossible for self-interested individuals to refrain voluntarily from their exploitative action.

At a practical level, this theory has created an understanding that the position of the state is crucial in managing the commons. This is because whichever of the three proposals is adopted, the government should first assume sole ownership over the resource. If the private ownership approach is accepted, the government will then divide the sea territory or resources into portions and distribute them to be controlled by fishermen or fishing companies. If the sole ownership approach is implemented, the government will grant a single

---

9   Sole ownership refers to a situation where an individual or entity owns the whole resource whereas under private ownership, individuals privately own different shares of the resource.

fisherman or fishing company a licence to control all fishing activity. For the third approach, the government will develop policies that constrain 'free exploitation' such as taxes, gear restrictions, quota systems, closed seasons and so on.

The opinions of Hardin and his supporters have been controversial, prompting worldwide debate on theoretical and practical levels (for example, Hardin and Baden 1977; McCay and Acheson 1987; Berkes 1989 and Pomeroy 1994) that have led to the emergence of an alternative view that recognises the significance of traditional communal marine ownership. The arguments raised in *The Question of the Commons* (McCay and Acheson 1987) critically examine the assumptions that lie behind the 'tragedy of the commons model'. The key assumptions are that: (1) common property means free and open access to all; (2) maximising self-interest is the logic driving resource use; and (3) no social norms function to regulate individual behaviour. These assumptions ignore the fact that the concept of property rights is subject to cultural difference. Cases discussed in this volume show that the commons does not necessarily imply a 'free for all' but rather refer to communal property rights that are defined in terms of local concepts about appropriate resource use. This means that there is a social group that claims ownership of the resources—and excludes others—and that among the owning group there are rules and norms which regulate who may exploit the resources and how, when and where they may be exploited (Ruddle and Akimichi 1984).

Furthermore, Balland and Platteau (1996: 176) have identified two crucial weaknesses to Hardins' proposed solutions regarding private property rights. The first weakness concerns the distributional effects of such arrangements. For example, when a single person or company controls a section of the sea, others are excluded from gaining benefits from the territory. Although the owner of the territory might employ them, any economic interest is likely to be much lower than if they share the same right of access to the resource. The second weakness argues that private ownership is no guarantee that the owner will use the resource efficiently or sustainably. Hence:

> [T]he possible sources of inefficiency of a privatization programme has [sic] been analysed by identifying the conditions under which the establishment of private property would automatically constitute an efficiency-increasing move compared with a regime of unregulated common property. More specially, property rights have to be well defined, all markets in the economy must exist and, moreover, be perfect and competitive, and there is no cost entailed in the enforcement of private property rights. In the real world, however, none of these

conditions is likely to be satisfied and, therefore, it can not be predicted a priori whether the establishment of private property over natural resource enhances efficiency or not (Balland and Platteau 1996: 176).

The proposed solution regarding centralised state resource management also comes under criticism by Balland and Platteau (1996) who point to a number of problems regarding the agendas, capacity, and functioning of states in developing countries. The problems include: limited human and financial resources needed to collect and analyse data on the condition of the resource; limited capacity to develop effective policy and regulation to monitor and enforce their implementation; subordination of environmental concerns; and resistance from resource users due to a lack of effective relations between local communities and state authorities. Under these conditions, development of sustainable and effective resource management systems is difficult to achieve.

The criticisms outlined have prompted widespread reconsideration of communal property rights associated with traditional communities of which communal marine tenure has been a focus. Anthropological studies of traditional marine tenure started gaining popularity in the 1970s (Ruddle and Akimichi 1984: 1) revealing that Hardin's notion that the sea is 'free for all' is not universally applicable. Some communities conceptualise the sea as subject to communal ownership whereby 'use rights for the resource are controlled by an identifiable group and ... there exist rules concerning who may use the resource, who is excluded from using the resource, and how the resource should be used' (Berkes 1989: 10). It is argued that the existence of communal property rights not only suggests that individual self-interest is not necessarily predominant in traditional communities, but that such communities have the ability to work together to create institutions which function to avoid the 'tragedy of the commons' and such forms of communal ownership sustain the equal distribution of the resources.

In particular, Berkes (1989: 11) highlights five important roles of communal property rights. First, they ensure livelihood security by enabling every member of a community to meet their basic needs through assured access to vital resources. The second role of community property rights is conflict resolution because these rights provide a mechanism for the equitable use of resources with a minimum of internal strife or conflict. Third, these rights serve to bind members of the community into a single compact unit because community property explicitly links group membership and resource control resulting in increased teamwork and cooperation. Fourth, communal property rights increase conservationism since they are usually based on the principle of 'taking what is needed' (ibid.: 12). Finally, communal property rights maintain ecological sustainability with communal management often incorporating ritual practices that synchronise resource exploitation with natural cycles.

However, evidence exists that the practice of traditional marine tenure and traditional marine resource management in general has diminished. Johannes (1978: 356) suggests that the market economy, the breakdown of traditional authority structures, and the imposition of new laws and practices by the state are key interrelated factors that have resulted in the degradation of traditional marine tenure in Oceania. Typically, when people participate in the market economy, money becomes central to their economic life. In an attempt to earn as much money as possible, marine exploitation intensifies through increased fishing time or the adoption of more effective fishing technologies. This trend is in line with government development policies based on profit maximisation principles. Under these circumstances, communities and governments apply pressure on traditional leaders to relinquish guardianship of marine resource management. The breakdown is compounded when colonial or modern governments introduce laws and other regulations based on the European tradition of the 'freedom of the seas' (ibid.: 358).

For Johannes and those who share his perspectives, such as Bailey and Zerner (1992), the erosion of traditional marine resource management is not only a loss of traditional wisdom but also a loss of the potential solution for the 'tragedy of the commons'. They argue strongly for formal government legislation to prevent complete dismantling of customary marine tenure through market pressures. Formal legal acknowledgment:

> will strengthen the ability of the owners to police their resources— something they often do voluntarily if their rights are secure. Legislation that weakens or nullifies marine tenure laws increases the government's regulatory responsibilities and places additional burdens on typically understaffed and underfunded fisheries department (Johannes 1978: 360).

Accordingly, appropriately designed government legalisation on customary marine tenure will preserve traditional practices, ensuring effective and sustainable resource management as well as reducing government responsibility regarding the crafting, monitoring and funding of marine management operations.

These suggestions by Johannes share similarities with the more recent proposal of the collaborative management (co-management) system of resource management. If one considers marine resource management as a continuum in which centralised government practice (government power) is at one end and community–based practice (fishermen's power) is at the other end, co-management lies closer to the fishermen's power (McCay 1995). In fact, 'the basic principle of co-management is self-governance but within a legal framework established by government, and power is shared between user groups and government' (McCay and Jentoft 1996: 239). This means that, unlike a centralised government system where all decisions are formulated and acted upon by the government (top-down), co-management involves fishermen in the decision-making process as well as in implementing management strategies. Pinkerton argues that:

> Basically, by instituting shared decision-making among these actors [government and user groups], co-management systems set up a game in which the pay-offs are greater for cooperation than for opposition and/or competition, a game in which the actors can learn to optimise their mutual good and plan co-operatively with long term horizons (Pinkerton 1989: 5).

Having the involvement of fishermen in the implementation of management plans—given the closed nature of the community and the close relationship between members of the community—is essential to support the effectiveness and efficiency of management practices especially where government capacity is weak. Thus, co-management creates the possibility for combining the strengths of the community with those of the government thereby eliminating the weaknesses of individual actors. Once it is established, 'the benefits sought by one or all of the actors through fisheries co-management are more appropriate, more efficient, and more equitable' (ibid.).

The answers to the question of how to create fishery co-management are various and debatable. However, there are two fundamental prerequisites for success: management responsibility needs to be delegated to the community (Pinkerton 1989); and relationships between stakeholders needs to be based on equality and developing dynamic partnership relationships between user groups (Nielsen and Vedsmend 1997: 55). McCay and Jentoft (1996: 239) suggest that the first prerequisite could 'result from a legal recognition of traditional, communal management and rights system'. In other words, to create effective and efficient co-management, the government should appreciate and acknowledge the existence of community management practices and rights. An acknowledgement would facilitate an equal balance of power between the government, local communities and other consumers such as the fishing industry. The acknowledgement of communal management and local rights serves to increase the bargaining power of the community in negotiation processes, particularly during the creation of management plans. Otherwise, the community is generally powerless compared to the government and the fishing industry. In the end, the acknowledgement of communal management and local rights might facilitate the creation of a fair negotiation process that avoids government and fishing industry co-opting of the community. In this way, the dynamic partnership between government, community, and other stakeholders may materialise and function to empower the community (Jentoft 2005).

Marine tenure is clearly central to the centralised, community-based and collaborative systems of resource management (Jentof et al. 1998). The notion of the sea as a 'commons' and the nature of man as 'rational individualistic' precipitated government involvement in managing marine territories and resources (Ostrom 1990). Likewise, the debate surrounding traditional marine tenure has generated support for the community-based resource management. For example, Jentof et al. (1998: 432) note that '[t]he argument for co-management is part of an attempt to recognise and buil[d] upon a larger set of property options for managing natural resources, including various forms of community-based jurisdiction over natural resources, or at least rights to use and manage them' (ibid.: 433). In this regard, the presence of communal property will make the creation of co-management easier.

In the Indonesian context, the Indonesian constitution states that maritime territory and resources are under state control. As a result, the Indonesian government has centralised the management of marine resources through the framing, implementing, and monitoring of related policies. However, the government lacks the resources required for the development of good fishery policies or for implementation and monitoring. An evaluation of government performance indicates economic rather than environmental interests drive fishery policies (Bailey 1988; Novaczek, Spacua et al. 2001) resulting in: over-exploitation; unequal economic distribution between small, medium and large-scale fishery operators, often leading to the marginalisation of local fishermen; and conflict between resource users (Bailey 1988; Bailey and Zerner 1992).

The discourse on *sasi* has emerged in response to the failure of centralised marine resource management. Initially, *sasi* was seen as a better alternative for marine resource management, ensuring more equal sharing of resources, preventing conflict between resource users, and contributing to resource sustainability. However, *sasi* is weakening, and in Maluku the practice has been largely abandoned, and the government has been called on to implement co-management. Harkes and Novaczek note that:

> *Sasi* thus offers a solid foundation for resource management in the region. It provides a structure that is culturally embedded, a functional enforcement mechanism, and a set of rules and regulations that are acceptable to most. The familiarity with management concepts, the acknowledgement of a need to protect natural resources, perceived benefits and general appreciation of sasi makes it highly legitimate. With a formally acknowledged *kewang* who have access to funds, training and a network, enforcement of regulations can be carried out locally in a legitimate way. With the assistance of NGOs, scientists and government, (co-) management structures could be established that include the principles and components of *sasi* (Harkes and Novaczek 2002: 257).

This quotation illustrates the underlying assumptions in the discourse of *sasi*, community-based resource management and co-management and how they all relate to the embedded nature of culture in traditional marine tenure and management. The three discourses highlight the attributes that such embedded socio-cultural norms have in influencing the effectiveness and efficiency of traditional marine tenure and management. In my experience, I find these assumptions questionable as the understanding of these traditional institutions and practices is only in the context of resource management. Examination of the socio-cultural context is based on observations that these institutions and practices have been implemented for generations, coordinated by traditional agencies, and guided by traditional values and norms. However, the question of what is the meaning of traditional marine tenure and resource management in a broader socio-cultural context has not been addressed with any rigour. The inter-relations between traditional marine tenure and management with other socio-cultural institutions and practices are barely examined.

This book aims to fill this gap by examining the social meaning of traditional marine tenure in a broader social context. It also looks at the relationships between traditional marine tenure and other traditional and modern institutions and practices. As previously stated, the meaning of traditional marine tenure is more complex than what is generally assumed in the discourse of resource management. Marine tenure is also an integral part of both the internal structure of a community and a larger structure of which the community is only a part. The relationship between marine tenure and contestation, and between 'the nobles' and 'the commoner,' as discussed in reference to Sather verses Tutrean conflict, is a clear example of these structurual dynamics. My arguments related to the raiding of an illegal fishing company, cyanide fishing, and clove season incidents are examples of the connections of marine tenure with both the internal structure of the community as well as government, military, law enforcement agents and fishing businessmen and women involved in these conflicts. It is evident that in these situations, traditional marine tenure serves more as a tool for social contestations than as an instrument of resource management. In fact, these conflicts confirm that marine tenure for resource management can be over-ridden by its social function which is contradictory to the stated goal of marine resource management, namely the sustainability of resources and the application of social justice in distribution of benefits.

The data discussed in this book was collected during field research in the Kei Islands of Southeastern Maluku, Indonesia from February 1996 to March 1997. Updates were obtained through a short visit to the Kei Islands in 2009, and by literature and news reviews. Although particular villages on Kei Kecil and Kei Besar have informed the discussion, the focus of study was the village of Dullah Laut on Kei Kecil and the villages of Tutrean and Sather on Kei Besar Island (Map

1-2). Dullah Laut is the village that has been researched most intensively and yielded a rich source of information that has facilitated a more comprehensive understanding and discussion of traditional resource management issues. Sather and Tutrean villages were studied less intensively but were of great interest due to their longstanding conflicts over marine territories, resources, and political leadership.

Structure of the Book

This book is divided into two main parts. The first part provides the overall setting and consists of four chapters. Chapter Two provides an overview of Kei Archipelago and its people in terms of traditional and modern structures. Chapter Three provides a detailed account of Dullah Laut Village, which forms the basis from which analysis is drawn regarding case studies discussed in the second part of this book. Unlike the first two chapters that discuss the general ethnography of the Kei Islands and Dullah Laut Village, Chapters Four and Five provide the detailed setting for the issue of marine tenure. Chapter Four discusses the position of marine tenure practice in traditional precedence contestation, and Chapter Five highlights the general characteristics of marine tenure practiced in the village compared with tenure practice in Watlaar Village on Kei Besar Island.

The second part of this book examines four cases of conflict over traditional marine territory. Chapter Six discusses marine tenure conflict related to the dynamics of local politics in the village. In essence, this chapter argues that in reference to the politics of the village head, control over marine territory is considered political capital rather than as an element of resource management. Chapter Seven discusses marine tenure conflict by considering its legal context. This chapter argues that the formal legal position of marine tenure is not so important given that at a practical level, people's behaviour is not governed by formal legal definitions but by considerations that are more pragmatic. The economic aspects of marine tenure are examined in Chapter Eight. My argument in this chapter is that while the market has eroded the practice of traditional marine resource management in other parts of Maluku, in the Kei Islands this is not the case. Market influences have enlivened traditional claims over sea territory.

The case discussed in Chapter Nine is that of Sather and Tutrean villages on Kei Besar Island. The main argument of this chapter is that even in the context of tradition, marine tenure is an object of dispute. The practice of marine tenure is an integral part of the whole social structure of the community and in fact, control over sea territory is a symbol of precedence. Therefore, when contestation over precedence arises, control of marine resources automatically becomes one of the contested issues.

The final chapter of this book offers concluding remarks. This chapter summarises my findings and examines their relevance in the contemporary search for more reliable systems of marine resource management. In this context, I situate my findings within the increasingly popular discourse of co-management.

2. The Kei Islands

The native boats that had come to meet us were three or four in number, containing in all about fifty men. They were long canoes, with the bow and stern rising up into a beak six or eight feet high, decorated with shells and waving plumes of cassowaries' hair. ... These Ké [Kei] men came up singing and shouting, dipping their paddles deep in the water and throwing up clouds of spray; as they approached nearer, they stood up in their canoes and increased their noise and gesticulations: and on coming alongside, without asking leave, and without moment's hesitation, the greater part of them scrambled up on our deck. ... These forty black, naked, mop-headed savages seemed intoxicated with joy and excitement. Not one of them could remain still for a moment (Wallace 1986: 420).

I was not welcomed with the same excitement on my first visit to the Kei Islands. This was not only because I shared similar skin colour to the people of Kei or that I landed at a different spot, but was more likely due to the people of the Kei Islands' familiarity with one and a half centuries of change since Wallace's visit in January 1857.

This chapter will highlight important changes that have occurred over the last century as well as introduce the Kei Islands and its people. A principal focus of this discussion will involve demographic and social changes in Kei society especially regarding population size, belief systems, and political structures. These changes have been fundamental in shaping the current political dynamic and in particular, the impact on the control and use of natural resources.

Location and Population

Location

The Kei Islands form an archipelago in the Arfura Sea, between 5°–6°5′ south and 131°50′–135°51′ east (see Map 1-2). They are located about 300–400 kilometres southeast from Ambon Island and the capital city of Maluku province. Discrepancies in island classification have resulted in Berhitu (1987) and KSMT (1993) classifying the Kei Archipelago with 100 islands, whereas Laksono (1990) claims there are 120 islands. Laksono (1990: 22) divides the islands into two major and three minor groups.[1] The Kei Kecil group is the first major group comprising two relatively big islands: Dullah (about 600 km[2]) and Kei Kecil (about 1300 km[2]).[2] Dullah is the home to Tual, the capital city of Southeastern Maluku Regency[3] and Ohoitel Village which was Laksono's 1990 research site. Despite its name meaning 'small', Kei Kecil is more than twice the size of Dullah and of particular significance in the region because it is where most of the government offices are located. The second major island group is Kei Besar. The main island, Kei Besar Island (585 km[2]) is the longest island in the archipelago and where Antunès had his research site in Watlaar Village in 1996/7. Kei Besar is surrounded by six small islands, only one of which is populated. The three minor groups are Kur, Tayando, and Tanimbar Kei, Barraud's 1979 research site.

At the time of Wallace's exploration in the mid-1800s, travel around the Kei Islands was limited to manpowered transport such as paddled or sailed canoes for short distances and the larger traditional Indonesian sailing boats, called sailed prau (*perahu layar*), were used for long distance voyages. Land transportation devices such as the bicycle were probably first introduced by Westerners in around 1882 when the Dutch established a remote area representative (*posthouder*) in Dullah Village, and Langen — a German— started a sawmill business in Tual.

Since the early 1990s weekly flights have operated from Ambon and more recently daily flights are available. In addition to commercial flights, a monthly military flight also operates to the Kei Islands. All air traffic uses the airport located at Langgur on Kei Kecil Island. Sea transportation has also developed greatly and probably well beyond the imagination of the people who boarded Wallace's vessel. Every fortnight, two big steel ships with hundreds of passengers arrive in Tual as well as smaller passenger and cargo ships, with an estimated 500 people coming and going from the Kei Islands fortnightly.

Motorised public transport in the form of small and medium-sized buses operates on Kei Kecil, Dullah, and central Kei Besar. Between the islands, there are two medium-sized wooden boats which transport people and goods from Watdek (on Kei Kecil) to Elat (on Kei Besar) twice a day. Other smaller wooden boats traverse the seas between outlying centres on a regular basis enabling

---

1   It is interesting to note the way in which earlier writers described these islands as it indicates the development of geographical organisation of the Kei Islands. Bezemer (1921: 229) classified the Kei Islands into four groups: Groot Kei (Kei Besar); Klein Kei (Kei Kecil); Tajando (Tayando); and Koer (Kur). GBNID (1944: 294) had a five group classification: Koer (Kur); Tajando (Tayando); Nehoerowa (Kei Kecil); Kai Dullah (Dullah); and Noehoetjoet (Kei Besar). Yet another division was proposed by the GBHD (1943: 148) consisting of Noehoe Tjoet (Kei Besar), Nohoe Rowa (Kei Kecil), Tajandoe (Tayando) Island, Koer (Kur), Keimeer (Kamer), Drie Gebroeders, Tengah and Boei. (Insufficient documentation exists to identify the last three of these islands.)
2   Dullah Laut and other smaller surrounding islands on the northern side between Dullah and Kei Kecil islands are part of this group.
3   In 2007, Tual became the centre of a new municipality bearing the same name. This municipality is the result of a split of the former Southeastern Maluku District. Tual municipality covers the island of Dullah and some outer islands of the Kei Archipeligo.

people to move between islands. However, despite the availability of motorised sea vessels, local sea transportation remains susceptible to the vagaries of the weather and small boats in particular are vulnerable to the strong currents and rough seas typical in the tropical archipelagoes. Remote islands such as Tam, Tayando, and Tanimbar Kei and the villages at the extreme northern and southern ends of Kei Besar are often difficult to reach due to bad weather. Even in the larger sailed prau, trips are dangerous during bad weather and islanders are reluctant to travel.

The advent of motorised transportation, both on land and sea, began sometime in the second half of the twentieth century and has been one of the key variables that intensified social interactions with the outside world. Unequal distribution and access to transport facilities can be seen as the major cause of uneven development which is illustrated by the high levels of development and infrastructure in the urban areas of Tual, Langgur and Watdek compared to the lack of development in remote villages and islands. Electricity is another modern invention that has impacted development in the Kei Islands. Regular power supply is only available in Kei Kecil, Dullah, and around Elat in the central region of Kei Besar Island and to a much more limited extent in villages where a number of small-scale residential generators operate. Access to power has enabled electronic communication and opportunities for increased interaction with outsiders in urban areas, but there has been limited change in villages without access to urban power supply.

Population

Collected census data is presented in Table 2-1. Population changes have been most notable over the last century. Although Wallace did not mention the population of the Kei Islands when he visited, Reidel (1886: 216) who was there shortly after, estimated that in 1882 the Kei Islands were home to about 17 246 people, 5580 of whom lived on Kei Besar and the rest populated Kei Kecil (5324), Tajando (665), Dullah Laut (391), Dullah (2352), Kur (1151), Kamer (195), Tam (790), Kei Tanimbar (322) and Hiniaar (476) islands. Comparing Reidel's estimation with the census conducted by the Dutch government in 1930, it seems that he had underestimated the population of Kei Besar which according to the latter was 25 229 people. Although the census was carried out 48 years later than Reidel's calculation, it is unlikely that the population of Kei Besar grew more than 350 per cent, while the population of Kei Kecil only increased by 13.9 per cent.

Table 2-1: Kei Kecil and Kei Besar population data (1882–2005).[4]

Subdistrict	Population census data					
	1882[a]	1930[b]	1971[c]	1990[c]	1995[c]	2000[d]
Kei Kecil	5324	13 289	60 616	67 507	73 729	78 902
Kei Besar	5580	25 229	33 730	38 820	39 473	41 249
Total	10 904	38 518	94 346	106 327	113 202	120 151

Sources: [a] Riedel (1886: 216); [b] DEZ (1936) [c] KSMT (1995: 33, 35); [d] BPSK (2003).

The census taken in 1971 indicates that the total population of the Kei Archipelago grew almost one and half times since 1930. In that period, the population of Kei Kecil increased more than three and half times while the population of the Kei Besar grew only slightly more than one-third (33.6 per cent). One possible explanation for this trend is that the period 1930–71 coincides with the development of the Kei Kecil as the administrative center of Maluku Tenggara District allowing Kei Kecil—together with Dullah—to become vibrant business centers in the region. Such developments have opened new opportunities, particularly in paid employment, which has drawn people to the area from Kei Besar and other rural islands of the Kei Archipelago. Migration from other districts in Maluku Province and even other provinces in Indonesia has also been evident.

Population increases were also noted for the census data from 1990, 1995, and 2000. From 1930–90, the population of the Kei Archipelago grew about 12.67 per cent accounting for 11.37 and 15.09 per cent growth of Kei Kecil and Kei Besar respectively. Between 1990-95 and 1995–2000, the total population rose by just over six per cent with the population of Kei Islands reaching 120 151 in the year 2000. As illustrated in Figure 2-1, population growth tended to rise slowly, except in the period of 1930-71. I suspect that the peak development which caused in-migration only took place from the 1960s to early 1970s, and the success of government family planning program introduced in the 1980s may also explain the slow growth trend.

---

4   After the application of *Local Government Law No. 22/1999* in 2000, Kei Besar and Kei Kecil islands subdistricts were split into five. The Kei Besar Subdistrict was re-organised into three subdistricts: Kei Besar; Kei Besar Utara; and Kei Besar Selatan respectively. The Kei Kecil islands sub-regency was restructured into Kei Kecil and Kur islands' subdistricts.

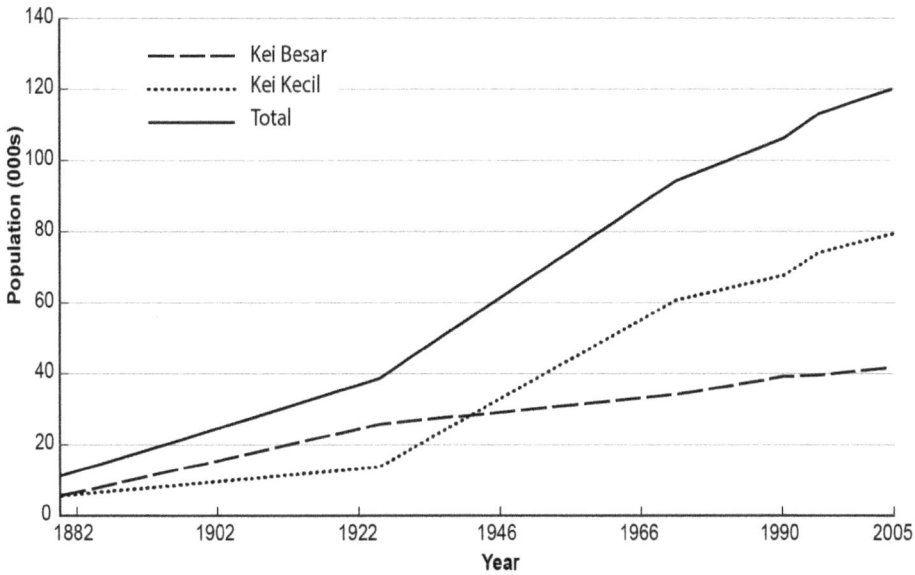

Figure 2-1: Kei Kecil and Kei Besar population growth (1882–2005).

Sources: Riedel (1886: 216); DEZ (1936); KSMT (1995: 33, 35); BPSK (2003).

Religion

The oldest known belief system in the Kei Islands was termed 'Alivuru' and has been described as a type of animism with the sun and moon gods as spiritual centre (Riedel 1886: 220; Van Hoëvell 1890; Barraud 1979, 1990a). A small number of villagers in Kei Tanimbar still acknowledge the Alivuru religion whereas others from this village call themselves Hindus.[5] However, some elements of belief and practice persist in most communities, regardless of assumed religion. For example, beliefs in ancestors or guardian spirits of sacred places—and the associated rituals—are still upheld by Protestant, Muslim, and Catholic communities.

Islam is the second oldest religion in the Kei Archipelago. Although the precise timing of the introduction to Islam is not known, Van Hoëvell (1890: 120) noted that there were no Muslims before 1864 but in the next decade, 30 per cent of the total population was of the Muslim faith. As in most other parts of

---

5   When the Indonesian New Order declared that there were only five official religions (Catholic, Protestant, Hindu, Buddhist and Islam), these people identified their religion as Hindu.

Indonesia, Islam was introduced through trade with Arabs, Makassarese, and Bugis. As traders settled on the islands, so the conversion to the Muslim faith began among the Kei islanders.

The introduction of Islam notably changed the settlement patterns among the Kei islanders, with converts to Islam geographically distancing themselves from pagans. Drawing from the contemporary situation, a reason for such separation could be the strict rules Muslims have in relation to food preparation. In particular, Muslims have strong concerns regarding food that may be contaminated by pork or the use of utensils that have been in contact with pork-related food. They also consider meat from animals not killed by Muslims as prohibited.[6] Increasing numbers of converts to Islam and the geographic isolation of Muslim communities has resulted in popularisation of the term 'Muslim settlement' (Ohoislam).

Catholicism was brought to the Kei Islands at the beginning of the twentieth century. The first Catholic mission in the Kei Islands was at the request of Adolph Langen to save the pagans from the domination of the Muslims.[7] Langen's views on the issue were shared by Van Hoëvell (1890) who was sent by the Dutch to verify his claims that the Muslims tended to exploit, manipulate, and introduce bad habits to the pagans (Van Höevell 1890; Schreurs 1992). Van Hoëvell noted that:

> [T]he Arabs and persons from Makasar do a lot of harm, they instil a hate against pagans and the European authority and also take advantage of their religious status. Everybody has to be submissive to the holy man and he has to have all he wants. The Keiese has to give his daughters, sell his produce to him at an unfair price and does not dare to complain (translated from Van Hoëvell 1890: 121).

They believed that adherence to the Islamic faith would jeopardise the good nature and character of the Keiese and felt strongly that conversion to Christianity would ensure the maintenance of their virtues and civilise the population. Driven by this belief, Langen sent a letter to the Central Catholic Mission in Batavia (now known as Jakarta) inviting their presence in the Kei Islands. The mission responded positively to this letter and on 12 October 1886, they sent a letter to the Dutch Governor General in Jakarta requesting permission for the establishment of a mission on the Kei Islands. The Dutch government consented and the first missionaries arrived in Tual in 1888 (Laksono 1990: 132; Renyaan 1996: 13).

---

6 Consequently, when other denominations host a ceremony with a mixed faith attendance, they will organise for Muslims to prepare the food for their Muslim guests.

7 Adolf Langen, a German who started a sawmill business in the Kei Islands in 1882, took an active interest in the social circumstances of the the Kei people.

The introduction of Catholicism and the establishment of a mission station at Langgur on Kei Kecil created tensions between the Muslim and Catholic converts. The tensions that developed were largely influenced by the mission's involvement in the political rivalry between segments of the community, as illustrated by the conflict that arose between the Muslim village of Tual and the new Christian village of Faan (see Map 1-2).

In the traditional context, Faan Village was under the control of Tual and as such, the village head of Faan was required to pay respect to the ruler of Tual. When Faan villagers converted to Christianity, the village head of Faan disputed the inferior position in relation to the rule of Tual and asked the mission to free them from the relationship. The missionaries supported this proposal since they saw that Islam was a significant challenge to their success in spreading Christianity. Despite some disagreement, the Dutch officials were finally persuaded and agreed to appoint the Faan village head as new ruler of the territory. The new appointment and order gave Catholics political superiority over Muslims and was a sign of political and cultural victory for the Dutch.

The Tual ruler and other Muslim leaders in Tual rejected the appointment. Assen, the Dutch representative in Tual, reported their protest:

> If there is a person who by their own, wants to be a Catholic, we will not hamper them, but if it will make a ruler or a de facto *orang kaya* ['wealthy person,' the title for a village head] withdraw their domain from a traditional bond, this means that the submission to a religion [Catholic] is a political rebellion against our power (translated from Schreurs 1992: 42).

The protest expressed a clear readiness for direct confrontation which was taken seriously by the Dutch. Thus, when they visited Tual, police men were brought along to protect them from possible Muslim attack. Apparently, this was a good policy to avert conflict between these two communities and open conflict never broke out between Faan and Tual.

Inter-religious relations were further complicated with the arrival of the Protestant Mission in the early twentieth century. The Catholic and Protestant missions were in direct conflict, converting as many pagans to their respective faith as possible and both sought support from the Dutch government. Laksono (1990: 145–56) describes an example of this conflict with the conversion of the pagan population of Ngat settlement on Kei Besar. The Ngat resisted the Catholic missionaries' efforts to convert them to Catholism and the mission responded with a request to the Dutch official in Tual to relocate the villagers to a neighbouring Catholic village called Bombay. The Catholic Mission believed that if the Ngat lived in the same village, they would eventually convert. The

Dutch official agreed and planned to move the people of Ngat. The leaders of Ngat however, contacted a Dutch Protestant minister in Tual and asked him to prevent the move and in return they would become Protestants. The conflict then became one between the Catholic and Protestant missions.

The Protestant Mission also became involved in conflicts between and within local communities. Lasomer (1985: 74, quoted in Laksono 1990: 149), describes a conflict that was triggered by a marriage arrangement involving communities living in the upper and lower areas of Soindat Village on Kei Besar (see Map 1-2). As a strategy to gain external support, the people living on the lower part of the village converted to Catholicism in 1909. In opposition, those from the upper part of the village converted to Protestantism. Subsequently, the two communities became separated by a marked boundary and connected to different external agencies for support. Disharmony was not only restricted to Catholics and Protestants but also characterised relations between Muslims and Protestants.

The Dutch government generally favoured Protestants and Catholics over Muslims. Referring to the situation in Ambon, Chauvel (1985, 1999) detailed how local Christians were assimilated into structures of colonial empire, gaining access to education and employment. In such positions as colonial government officials and military officers, they straddled ruling in the name of the colonial power on the one hand, and being ruled by their colonial masters on the other. In this situation, it was not uncommon for some of them to believe that they were superior to others in the community.

When the Japanese reached the Kei Islands in 1942, the situation was reversed (Chauvel 1985, 1999). The Japanese removed all of the structures developed by the Dutch, including the Protestant and Catholic missions. Thirteen Catholic leaders, including the bishop working at the mission, were executed by Japanese troops shortly after landing in Tual. Many Protestants associated with the Dutch were also killed and some churches were destroyed.

In these new circumstances, both Catholics and Protestants feared Muslims. Father Bedaux (1978), who escaped the massacre because he was working in a village some distance from Tual when the Japanese troops raided the mission in Langgur, openly expressed his and his followers' fear of and suspicion toward the Muslims. Informed that the mission leaders had been killed and the Japanese troops were looking for other priests, he ran from the village. During his escape, he avoided passing by Muslim settlements or greeting Muslim villagers. This was driven by the common suspicion held by Catholics that Muslims were Japanese spies and that their sightings would be reported to the Japanese troops. These suspicions were confirmed by Lawalata (1969: 32–5) who recorded that the Japanese did recruit Muslims to spy for them and work at the Japanese administration office.

The disharmony between religious groups has continued and although relations appear cordial, there remains a palpable uneasiness between them. This has been highlighted by the more recent conflicts in Central and North Maluku Province and even in the Kei Islands. Conflicts between Muslims and Christians which started in Ambon in January 1999, lasted for years resulting in more than 3 000 deaths, forcing more than 123 000 to become refugees, and destroying billions of rupiah in property (see Van Klinken 2001 and International Crisis Group (ICG) 2002). In the Kei Islands, similar conflicts took place in 1999. Two of the bloodiest incidents as well as several minor clashes that broke out in that year caused the deaths of 200 people, injured hundreds, destroyed around 4 000 buildings, leveled more than 20 villages, and internally displaced around 30 000 people. (Topatimasang 2004: viii). Although the violence has ceased, it is not impossible that similar incidents could erupt in the future.

The combined impact of religious and political ideologies constructed through conflicting circumstances and the spatial distance between these communities has added fuel to these conflicts. As with the Muslims, the Catholics and Protestants have created separate settlements that have resulted in distancing genealogically connected individuals and constrained the development of avenues for communication. It is evident that these divisions are becoming more entrenched and isolating family members. Some provocations could instigate violent clashes, as in the above mentioned conflicts.

A breakdown of the Kei population based on religious following over the last century is presented in Table 2-2. This data clearly shows that the founding belief system of paganism has been almost entirely lost from the population. Although the Muslims have always outnumbered the Catholics and Protestants separately, collectively the Christians have outnumbered the Muslims since sometime in the 1920s. Comparing Christian groups, there have always been more Catholics than Protestants and more recently it appears Catholic numbers are increasing more so than Protestant numbers.

Table 2-2: Beliefs systems followed in the Kei Islands (1887–2000).

Belief	Census/survey year			
	1887	1915	1930	2000
Muslim	5893	12 000	20 000	50 242
Protestant	...	3000	11 000	28 663
Catholic	...	8000	13 000	41 138
Other	14 137[a]	7000[a]	6000[a]	108[b]
Total	20 030	30 000	50 000	120 151

Note: [a] Pagan, [b] Hindu, Buddhist and Pagan.

Sources: Taken from Laksono (1990: 26, 122) and BPSK (2000).

Political Organisation

Understanding the dynamics of political organisation in the Kei Islands revolves around two ideals. First, political organisations frame the social relations between people and groups, making it an important source of power. The position of village head, for example, is a source of power that enables this person to make decisions about the lives of so many villagers. Second, Kei people consider change from one political organisation to another more as a process of enrichment or accumulation rather than replacement of an old by a new type of organisation. So, when external powers such as the colonial or Indonesian government introduced different types of social organisations, they added more avenues from which to source power which complicated issues of contestation and conflict. The case studies in later chapters will give concrete examples of these complexities, while this section will only describe the structural dynamics of political organisations. In this regard, the discussion is divided into two parts—the first part relates to traditional political organisations, and the second deals with 'modern' political organisations.

Traditional Political Organisation

The traditional political organisations in the Kei Islands are hierarchically ordered with the overarching organisation refered to as a *lor* (literally meaning 'whale'). There after the rankings in descending order are: kingdom (*utan*, *ratschap*);[8] village (*ohoi*); and hamlet or settlement (*ohoi kot*).[9] In this system of organisation, the smaller groups form an integral part of the larger organisation (Figure 2-2).

Categorically, Kei islanders are divided into three *lor*, referred to as 'nine group' (*lor siw*), 'five group' (*lor lim*) and 'neutral group' (*lor labay*). Membership of a particular social group within one of the three *lor* is based on the narrative of group formation, which starts with the appointment of a king (*rat*).

---

8 Most if not all *utan* were converted to political units called 'kingships' (*ratschap*) during the Dutch colonial period. Now the term *ratschap* is more popularly used than the Kei term *utan*, but in rituals people still use the vernacular term.
9 Actually, there is no special term in Kei language that refers to this political unit. In a District Regulation on Ratchap (spelled as 'ratshap') and Ohoi issued in 2009, this political unit is also called 'ohoi'. I asked the ex-village head of Dullah Laut, who had become a member of Tual City Legislative Assembly (Dewan Perwakilan Rakyat Daerah) when I met him in 2009, to differentiate between the two and he suggested adding the word '*kot*' which literally means 'small'.

Lor level Led by primus inter pares of the kings	Lor Lim (5 group)	Lor Labay (neutral group)	Lor siw (9 group)
Kingdom level (utan/ratscap) Led by the king (rat)	Ten utan (distributed on Kei Besar, Kei Kecil and Mangur islands)	Two utan (Tam on Tam Island and Werka on Kei Besar Island)	Ten utan (distributed on Kei Besar, Kei Kecil, Dullah and Kamer islands)
Village level (ohoi) Led by the traditional village leader (orang kaya)	more than 100 ohoi (each affiliated to an utan)	one ohoi in each utan	More than 100 ohoi (each affiliated to an utan)
Settlement level (ohoi kot) Led by settlement leader (bapak soa)	Some ohoi have more than one ohoi kot, some others do not as the ohoi form a single undivided political unit		

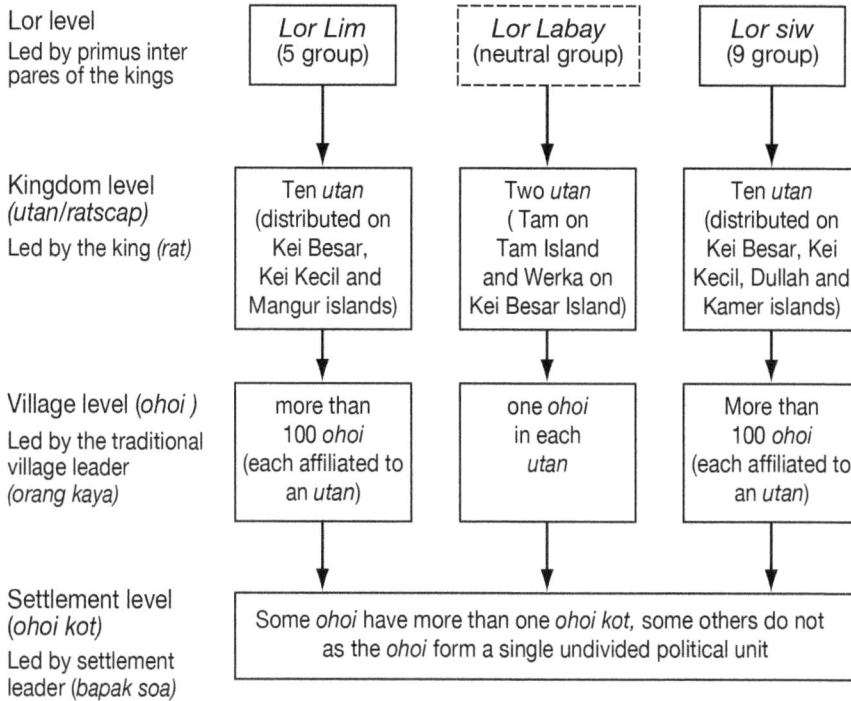

Figure 2-2: Traditional political structures used in the Kei Islands.

Note: The dashed line represents a different organisational form compared to the other *lor*.

Source: Author's fieldwork.

The appointment of a king determines whether his kingdom would be a member of the nine, five, or neutral group. The membership of a social group is taken from membership of a particular kingdom. Thus, in determining what *lor* a village belongs to, people first consider which kingdom the village is attached to and then determine the membership in relation to the nine, five, or neutral group. For example, in determining whether the villagers of Dullah Laut belong to the nine group, five group, or the neutral group, they will indicate that their village is a member of ' the three villages in the west' (*lor utan til warat*) led by King Baldu of Dullah village. Based on this, they claim membership of the nine group since this kingdom is a member of the nine group.

The first two groups (nine and five) are often associated with the moiety system, a system that divided the community into two distinct but complementary groups, which was common in eastern Indonesia. In Maluku, this system 'once encompassed all societies from Seram to Aru in the southeast and that even now is ideologically important' (Valeri 1989: 117).[10] Since each group in the

---

10   Seram is an island north of Ambon while Aru is an archipelago to the southeast of Kei (see Map 1-1).

system named themselves using a numerical index nine (*siwa*) and five (*lima*), Cooley (1962) and Valeri (1989) called them as 'nine' and 'five' moieties. They also argued that this division was associated with symbolic marks such as (1) 'autochthonous' verses 'immigrant, and (2) 'landward' verses 'seaward'. The first mark referred to the association of 'nine' or 'five' groups either as autochthonous or immigrant population, while the latter mark pointed to the association of either group to their coming, settlement, or territorial control.

Interestingly, despite the same numerical indices—nine and five—these two groups have different characteristics from the nine and five moieties described by Valeri (1989) and Cooley (1962). By way of clarification, we can examine the application of symbolic marks associated with nine and five groups, such as autochthonous, immigrant, landward and seaward. Considering the narratives of the nine and five group, the association of the two groups with autochthonous and immigrant marks do not apply because the founding fathers of both groups were immigrants and because the narratives note that the recruitment of group membership was conducted by appointing a king in each domain. This appointment meant that the king's domain became members of the nine or five group. The issue of political leadership in the Kei Islands that privileges the nobles who are almost all—if not entirely—immigrants, is discussed in detail in Chapter Four.

The distinction between the nine group and five group as it relates to the seaward and landward marks is not applicable either for two reasons. First, almost all villages in the Kei Islands are located in coastal areas where there is no conceptual distinction between those who live on the landward or seaward positions. Second, the distinction between the geographical distribution of nine group and five group members is not particularly relevant since both are distributed evenly throughout the archipelago (Map 2-1).

However, the character of rivalry between the nine and five group was shared with their corresponding moeties. In fact, one of the most important factors that drove each group to recruit as many members as possible was to defeat the opposing group or, at least to strengthen their defense against attack. In essence, it was a 'system' of alliance to accumulate strength to overpower the opposing group. Renyaan (1990: 33) noted that there were 11 wars between the five and nine groups which concluded when the two groups agreed to end hostilities.

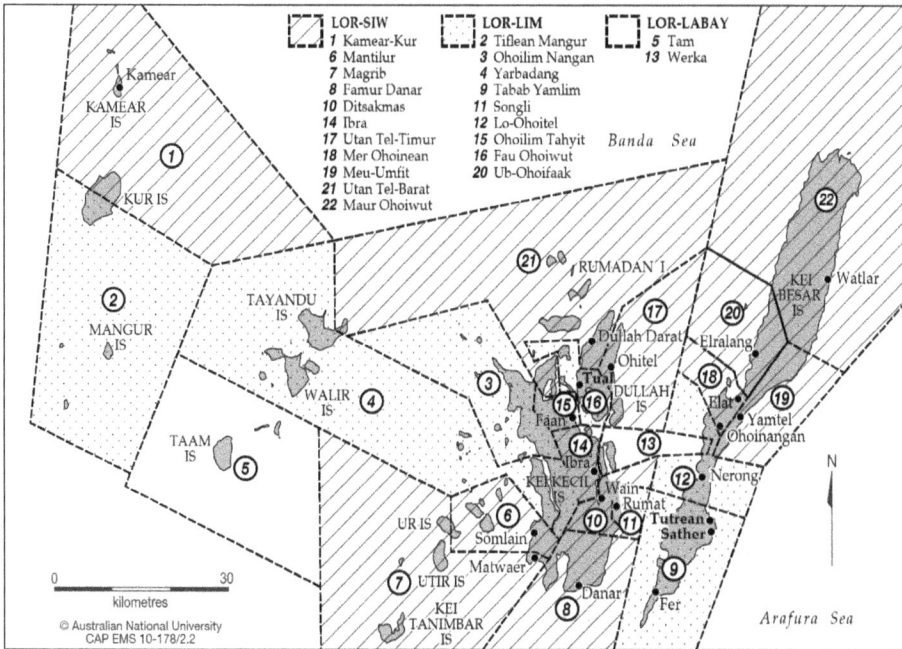

Map 2-1: Spatial distribution of *Lor Siw*, *Lor Lim*, and *Lor Labay*.

Source: Modified from Rahail (1995).

The narratives of group formation also determined the position of a particular member of the nine or five group in the organisation of the group. Arnuhu and Bomav were considered founding fathers of the nine and five group respectively. The seniority of other members was taken into account in the recruitment process and those who were recruited earlier were considered a 'big brother' to those who were recruited later. In theory, senior members hold more power than junior members, however in practice the position of those with more power was a constant source of contestation. In fact, Van Hoëvell (1890: 123) notes that the nine group has had many leaders— first the king from Wain, later the ruler from Danar, and finally the ruler from Dullah.[11]

The discussion of the moiety system of the nine and five group does not focus on the neutral kingdom (*lor labay*). From the narrative describing recruitment of the nine and five group, it is evident that the neutral kingdom groups which consist of Tam on Kei Kecil and Werka on Kei Besar were neglected in the recruitment process. As these two kingdoms did not coordinate as allies or as a distinct social group, it is difficult to ascertain whether the kingdoms formed a single organisation similar to the other two groups. However, these kingdoms

---

11   This version is different from the narrative of origin which mentions that the King of Danar was the founding father of the nine group (see also references to this in Chapters Three and Four).

were also called 'mediators' or 'neutral' groups.[12] These names indicate that the neutral kingdom might play an important role in encouraging the two conflicting *lor* to reach an agreement, although there was no evidence of this in the literature.

The function of *lor* changed over time. In the ancient time, as indicated by the narratives, the *lor* functioned as an organisation of allies and played an important role in settling conflicts between its members. In times of war those belonging to one *lor* would help fight another. However in modern times, rather than fighting when there is conflict, the nine group, five group, and neutral kingdom are more likely to work together to resolve customary problems that they cannot solve alone. In such cases, a committee of leaders from both five and nine groups and the neutral kingdoms is established. An example of such a situation was the committee that formed for solving the dispute between Tutrean and Sather villages in 1990 (detailed in Chapter Nine).

In daily life however, these groupings are not so important. Although elders still remember the narratives and key events relating to the *lor*, they do not see it as a significant element that directs their daily life. The Sather villagers' rejection of the decision made by the committee consisting of leaders from the nine and five groups and the neutral kingdom was an example of this. Yet, sometimes lingering tensions become known. For example, during the celebration of the fiftieth Indonesian Independence Day, the two groups competed in the wooden war boat race and a fight broke out among them because one of the rowers in the winning boat was not a member of the boat's *lor*.

The kingdom (*utan* or *ratchaap*) designated during colonial rule is the second highest traditional level of political organisation. There are 22 kingdoms in the Kei Islands— ten in each of the nine and five groups, and two kingdoms categorised as neutral (see Table 2-3). A ruler or king (*rat*) led each kingdom. In organising the territory and people, he was assisted by prominent leaders in his domain including village heads under his control. In difficult times, the king would arrange a meeting attended by all leaders of the village and other prominent leaders in the kingdom. The king would lead the meeting and discuss any issues in need of resolving.

---

12   The use of the words neutral or *penengah* instead of vernacular terms might indicate that the meaning of *lor labay* as 'mediator' is a modern conception.

Table 2-3: Membership of *Lor Siw*, *Lor Lim*, and *Lor Labay*.

Lor	First ruler	Political domain	Location	Village centre
Lor Siw				
1	Arnuhu	Famur Danar	Kei Kecil	Danar
2	Matan Vuun Sutra	Ditsakmas	Kei Kecil	Wain
3	Magrib	Magrib	Kei Kecil	Matwair
4	Baldu	Utan Tel Warat	Dullah	Dullah Darat
5	Airaar Vavav	Utan Tel Timur	Dullah	Ohitel
6	Tali Larwai	Mantilur Kasilwut	Kei Kecil	Somlain
7	Elkel	Meu-Umfit	Kei Besar	Yamtel
8	Bar Vav Tanlain	Maur Ohoiwut	Kei Besar	Watlaar
9	Ohinangan	Ohinangan	Kei Besar	Ohinangan
10	Kilmas	Kamer-Kur	Kamer	Kamear
Lor Lim				
11	Bomav	Tabab Yam Lim	Kei Besar	Fer
12	Ihibes	Lo-Ohotel	Kei Besar	Nerong
13	Kirkes	Ibra	Kei Kecil	Ibra
14	Bal-bal Faan	Ohoilim Tahit	Kei Kecil	Faan
15	Yarbadang	Yarbadang	Kei Kecil	Tetoat
16	Songli	Songli	Kei Kecil	Rumat
17	Balaha Rahawarin[a]	Ub Ohoi Fak	Kei Besar	Elralang,Mar, Wer, Uwat[b]
18	Tufle	Tual	Kei Kecil	Tual
19	Rumadian	Ohoilim Nangan	Kei Kecil	Rumadian
20	...	Tiflean Mangur	Mangur	Tiflean
Lor Labay				
21	Taam	...	Tam	Taam
22	Werka	...	Kei Besar	Werka

Note: [a] rat from Elralang, [b] alternately.

Source: Adapted from Rahail (1995).

The incorporation of different village heads into a kingdom was based on narratives of the kingdom's formation. Since there are 22 kingdoms, there are at least 22 narratives recounting each kingdom's formation. However, each narrative explains the connection of a particular village leader to a king. Villages where the head is associated with a particular king become members of the kingdom, but this was all —including the position of king—subject to contestation. Van Höevell (1889: 108) noted that Har and Mun used to be under the king of Watlaar but they broke the connection declaring themselves free from the king's domain. The village of Langgiar also tried to break away from the control of the king of Fer, but they failed.

The village (*ohoi*) is the political organisation that is smaller than that of the kingdom and is led by a village head referred to as a 'wealthy person' (*orang kaya*) or in a few cases, by a military commander (*kapitan*). The village head is assisted by other leaders in the village such as an imam (Muslim religious leader), landlord (*metu duan* or *tuan tan*), war commander (*panglima perang*), and so on. The smallest political unit is the settlement (*ohoi kot*) which is led by 'the father of the settlement' (*bapak soa*).

In summarising the practical function of the traditional political organisation in the Kei Islands, the settlement can be viewed as the most important political organisation because it was integral to people's daily lives and handled problems that occurred in the community. Once the settlement resolved an issue, there would be no further need for discussion on the matter. Only if the settlement organisation could not resolve an issue would the problem be brought to the village level. In such situations, the village head and his village functionaries would be called on to assist and this pattern continued up to the *lor* level until the issue was resolved satisfactorily.

'Modern' Political Organisation

The introduction of modern political organisation in the Kei Archipelago dates back to the period of the Dutch occupation in Maluku. Contact with the Vereenigde Oost-Indische Companie (VOC)[13] in 1622 can be considered the starting point toward the incorporation of the Kei Archipelago into the Dutch East-Indies and the origins of 'modern' political organisation. At this time, the VOC signed an agreement with the leaders of Har, Laar, and Add villages on Kei Besar Island and installed its representative in Elat before 1636 (Reidel 1886: 218). In 1661 and 1664, villagers on Kur Island prepared agreements with the VOC and Riedel concluded that these contracts placed the people under their control (ibid.). The incorporation of the Kei political organisation into the 'modern' bureaucracy however, did not occur until after 1816 when the Dutch took control of Maluku from the British. At this time, Dutch officials began legalising the appointment of leaders and sorting out reported problems during intermittent visits to the Kei Islands.

Dutch control over the Kei people was established two centuries after the first contact between the people and the VOC. In 1882, the Dutch governor in Ambon set up a remote area representative (*posthouder*) in Dullah Village on Dullah Island. In 1892, the status was raised to that of sub-department (*onderafdeling*)

---

13   Before the imposition of colonialism in the nineteenth century, the United Dutch East Indies Company (VOC) controlled portions of the Dutch East Indies (the future Indonesia) during the seventeenth and eighteenth centuries. The VOC was a chartered company that owned a powerful naval fleet and employed European and native soldiers.

led by a domestic administrator (*controleur*) in Tual (Laksono 1990: 137). This change put the position of the Kei Islands in the Dutch political organisation as follows:

1. Maluku was considered a province, led by a governor based in Ambon.

2. The province was divided into several residencies, each of which was led by a resident.

3. The resident controlled several departments, each headed by an assistant resident. One of the departments was the South Islands Department with its office in Tual.

4. Every department consisted of several 'sub-departments', each run by a domestic administrator. The South Islands Department consisted of sub-departments based on Tanimbar, Babar, Kei, Aru, and Kisar islands.

In light of this structure, it is apparent that the Dutch incorporation of institutions in the Kei Islands did not change the traditional political structure because the unit of the Dutch political organisation that covered all of the Kei Islands acted principally as mediators rather than as 'judges' when dealing with villagers' issues. The traditional political organisation was effectively 'underneath' the Kei Islands sub-departments. This is best illustrated by the commission established by the Dutch to deal with community conflicts called 'the Great Council of Leaders in the Kei Islands' (*Groote Raat van Hoofden der Kei-Eilanden*) which consisted of prominent local rulers. Dutch decisions were mostly based on consultation with this committee (this example is discussed in Chapter Nine).

Nevertheless, we should also note that the Dutch had vested interests. As a consequence, even though their decisions were based on consultations with traditional leaders or their understanding of local tradition, fulfilling Dutch interests was their first priority. Therefore, policies were not always supportive of tradition. By way of illustration, if the Dutch had been effective mediators at appointing the ruler in Faan, they would have consulted the ruler of Tual or discussed the matter with the great council of leaders. It would appear that their interest in demonstrating the superiority of Christianity—which was associated with Dutch civilisation—over Islam was in conflict with their role as mediator.

Dutch policies also tended to be based on Western conceptions of 'state' and governance. The consequence was that the application of Western concepts to the traditional political organization and structure lead to conflicts. When the Dutch formalised the position of a particular person as a village leader, for example, they treated him as a village leader according to Western rather than traditional conception. The Dutch understanding of a village head presumed that he assumed all power related to the social and territorial organisation of the village. The traditional understanding is that the village head controls the political domain only, while issues of territory are under the control of the landlord. Or at least, the traditional view was that political and territorial control were two different issues and control over these issues was subject to contestation between different segments of the community (this point is elaborated on in Chapter Four).

The Dutch administrative position was also more highly recognised than that of the traditional political position, the consequence being that the policies of the Dutch could be used by local communities, or community factions, as a source of legitimacy. Because Dutch decisions were not always in accord with tradition, and tradition itself was subject to different interpretations, those who were favoured by the Dutch involvement in local issues would use their interpretations as additional 'ammunition' against their rivals. In other words, Dutch involvement made it possible for particular segments of communities to say, 'look, even the Dutch define me or my group as right. That means you're wrong'.

Another significant structural change to the political organisation occurred in the 1950s when the Kei Islands were declared part of the Republic of Indonesia.[14] Under Indonesian law and regulations, the political organisation of the Kei Islands could be summarised as follows and illustrated in Figure 2-3.

1. At a provincial level, the Kei Islands are part of the province of Maluku, headed by a governor in Ambon.

2. Maluku province consists of several districts, one of which was the district of Maluku Tenggara, led by a head of district in Tual. In 2007, Tual was separated administratively at the district level and called a municipality, resulting in the Kei Archipeligo now having two district administrations.

3. The district is divided into several subdistricts, each under the leadership of a subdistrict head. Until 2000, the Kei Islands were incorporated into two subdistricts— Pulau-pulau Kei Kecil, and Pulau-pulau Kei Besar. The office of Kei Kecil subdistrict was in Watdek on Kei Kecil Island and the Kei Besar subdistrict office was in Elat, Kei Besar Island. From 2001, these subdistricts split further with Kei Besar Islands now organised into three subdistricts: Kei Besar; Kei Besar Utara; and Kei Besar Selatan respectively. Kei Kecil Subdistrict has now been divided into seven subdistricts: Dullah Selatan; Dullah Utara; Pulau Tayanto-Tam; Pulau-pulau Kur; Kei Kecil; Kei

---

14   Changes due to Japanese influence are not detailed here as the impact on local institutions was not as apparent.

Kecil Timur; and Kei Kecil Barat. The first four subdistricts are under the Tual Regency and the rest are parts of Maluku Tenggara District.[15]

4. Finally, every subdistrict controls the smallest political unit called the modern village (*desa*).[16]

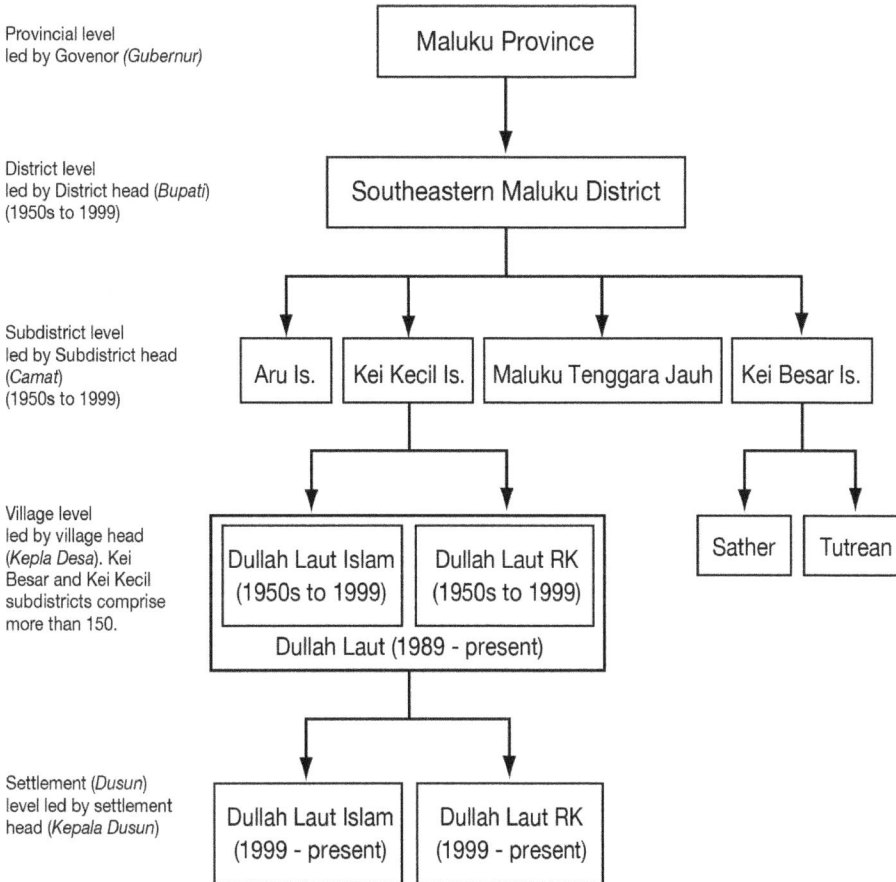

Provincial level
led by Govenor *(Gubernur)*

District level
led by District head *(Bupati)*
(1950s to 1999)

Subdistrict level
led by Subdistrict head
(Camat)
(1950s to 1999)

Village level
led by village head
(Kepla Desa). Kei
Besar and Kei Kecil
subdistricts comprise
more than 150.

Settlement *(Dusun)*
level led by settlement
head *(Kepala Dusun)*

```
                        ┌──────────────────────┐
                        │   Maluku Province    │
                        └──────────┬───────────┘
                                   │
                        ┌──────────▼───────────────────┐
                        │ Southeastern Maluku District │
                        └──────────────────────────────┘

  ┌─────────┐  ┌──────────────┐  ┌──────────────────────┐  ┌──────────────┐
  │ Aru Is. │  │ Kei Kecil Is.│  │ Maluku Tenggara Jauh │  │ Kei Besar Is.│
  └─────────┘  └──────────────┘  └──────────────────────┘  └──────────────┘

  ┌────────────────────┬────────────────────┐     ┌─────────┬──────────┐
  │  Dullah Laut Islam │   Dullah Laut RK   │     │ Sather  │ Tutrean  │
  │  (1950s to 1999)   │   (1950s to 1999)  │     └─────────┴──────────┘
  ├────────────────────┴────────────────────┤
  │      Dullah Laut (1989 - present)        │
  └──────────────────────────────────────────┘

  ┌────────────────────┬────────────────────┐
  │  Dullah Laut Islam │   Dullah Laut RK   │
  │  (1999 - present)  │   (1999 - present) │
  └────────────────────┴────────────────────┘
```

Figure 2-3: Keiese positions within the Indonesian political structure.

Source: Author's fieldwork.

Unlike the Dutch, the Indonesian political organisation overlaps with the traditional political organisation, down to the village level. Since Indonesian laws and regulations assume that this government-created political organisation should replace the traditional political organisation, the people's political organisation formally changed. The kingdom of villages was no longer

---

15   In 2007, ICG produced an interesting briefing on the political process of this separation that looks at the political maneuvers of local leaders to both support and oppose the separation of the district and subdistricts.
16   The term used for village political unit is different from one law to another (see Chapter Six). I added an example of Dullah Laut village in the figure as a reference to the discussion of Chapter Six.

coordinated by a king, instead they were under the control of a subdistrict head (*camat*) and the former 'traditional village' (*ohoi*) was known as the 'modern village' (*desa*).[17] At the village level, the application of the Indonesian political system also created problems because a traditional village did not automatically convert to a single modern village. In order to attract a larger central government subsidy—which was based on the number of modern villages—some traditional settlements (*ohoi kot*) in a traditional village were converted to a modern village.[18] That meant that the traditional settlement head (*kepala soa*) who was formerly under the coordination of a traditional village head (*orang kaya*), came under direct control of a subdistrict head because his traditional settlement was converted to a modern village and he became a modern village headman (*kepala desa*).[19]

To conclude this section, I would like to emphasise that at a practical level, people viewed these structural changes as cumulative, not consecutive. This meant that in a particular context they might use a structure that was not formally applicable and in other contexts, they might use more than one structure. Some of the cases discussed in the following chapters will illustrate this tendency.

Conclusion

Throughout their history the Kei people have been exposed to different structural arrangements. In terms of religion, the Kei people were exposed to Islamic, Catholic and Protestant religions. Once they chose to be part of a particular structure, this put them in opposition to others who adopted other religious structures. In terms of politics, the Kei also experienced the formal structural changes introduced by the Dutch and Indonesian states which are not compatible with their traditional political structures.

Interestingly, these structural encounters and changes were not thought to be replacing old structures with new ones. It seems that they considered these changes more as a process of enrichment of their structural preferences. For

---

17   With this introduction of a different type of village organisation, *utan* will from now on refer to 'traditional village' and *desa* refers to 'modern village'.

18   In his report, Berhitu (1987) explained that in order to absorb more government subsidies, 67 traditional settlements were converted to a modern village in 1970. When the Village Government law was applied, 44 traditional villages became 111 modern villages.

19   In 2009, the district of Maluku Tenggara passed regulations on Ratschaap and Ohoi that attempted to revitalise the traditional political organisation/structure. This was a response to the decentralization processes that have taken place since the collapse of the New Order Regime in 1998. Interestingly, they also adopted some ideas on 'modern' village government such as the requirements for the *orang kaya* candidate to be loyal to Pancasila (Indonesian Five Ideological Foundations) and to have graduated from at least the senior high school level. These regulations are another example that people do not consider the new structural changes to be replacements of the old. In fact, these regulations show that people produce and re-produce structure using either tradition or 'modern' elements in response to the contextual changes.

example, although Kei people are formally, Muslims, Catholics, or Protestants, they still practice rituals pertaining to their old religion. Despite the fact that their formal political structure should be the Indonesian 'system,' people still resolve their problems with customary law and procedures.

This situation could have advantages. If the Kei people found that a particular structure was incapable of defining or sorting out a problem, they still had other choices. However in a conflict situation, this circumstance can complicate and worsen the conflict since the conflicting parties might use different, incompatible structures in trying to legitimise their actions or claims.

3. Dullah Laut

Geography

The territory of Dullah Laut includes a group of islands that lie to the northwest of Dullah Island (see Map 1-2).[1] The islands that comprise this territory include: Dullah Laut (Duroa); Moa; Adranan (Dranan); Rumadan Warwahan; Rumadan Warohoi; Sua; Baer; Ohoimas; and Watlora (Ruin) (Map 3-1). The main island of the territory, Dullah Laut, is the largest of the group with an area of 8.1 km². Except for Rumadan Warwahan, Baer, and Ohoimas Islands, each of which is about 1.35 km² in area, each island is somewhat different in size. Rumadan Warohoi, at approximately 2.34 km², is the second largest. Adranan is the smallest being less than one square kilometre in area.

The geological conditions of these islands are similar to those generally found in the Kei Archipelago: low-lying limestone islands covered with a very thin layer of soil. This poses significant constraints in terms of access to a potable water supply. A small lake on Dullah Laut Island that could potentially provide a good supply of fresh water if it was on a non-limestone island has salt water permeating through the lime stone pores, and the height of water in the lake ebbs and flows with the tide. Dullah Laut Island is the only one of the nine islands that has limited fresh water available. The water has been primarily obtained from wells that are grouped into two sets, one set is used by the people of Ohoislam and the other by the people of Ohoisaran mainly for drinking purposes and washing. During the dry season, the waters of these wells become too salty to use forcing people to find fresh water elsewhere. They mostly obtain it from Dullah Island.

According to a survey conducted by a team from the University of Pattimura in Ambon, most of these islands are covered by fields with mixed annual crops, coconut plantations, and secondary forest woodland and shrubland. The densest mixed crops are found on Dullah Laut Island. In the early twentieth century, Rumadan Warwahan, Rumadan Warohoi, and Ohoimas Islands were cultivated intensively but these islands were abandoned during the Second World War and the islanders were forced to move to Dullah Laut Island for their safety, making it easier for the colonial government to exercise control. In the last two decades, there has been a gradual return to these islands and new fields have been planted. Coconut trees line the coastal areas of the islands. Apart from coconut plantations and new fields, the islands have a cover of secondary forest with woodland or scrubland.

---

1   The actual number of islands in this territory is contested since a kin group in Rumadan village claims ownership over both Rumadan islands.

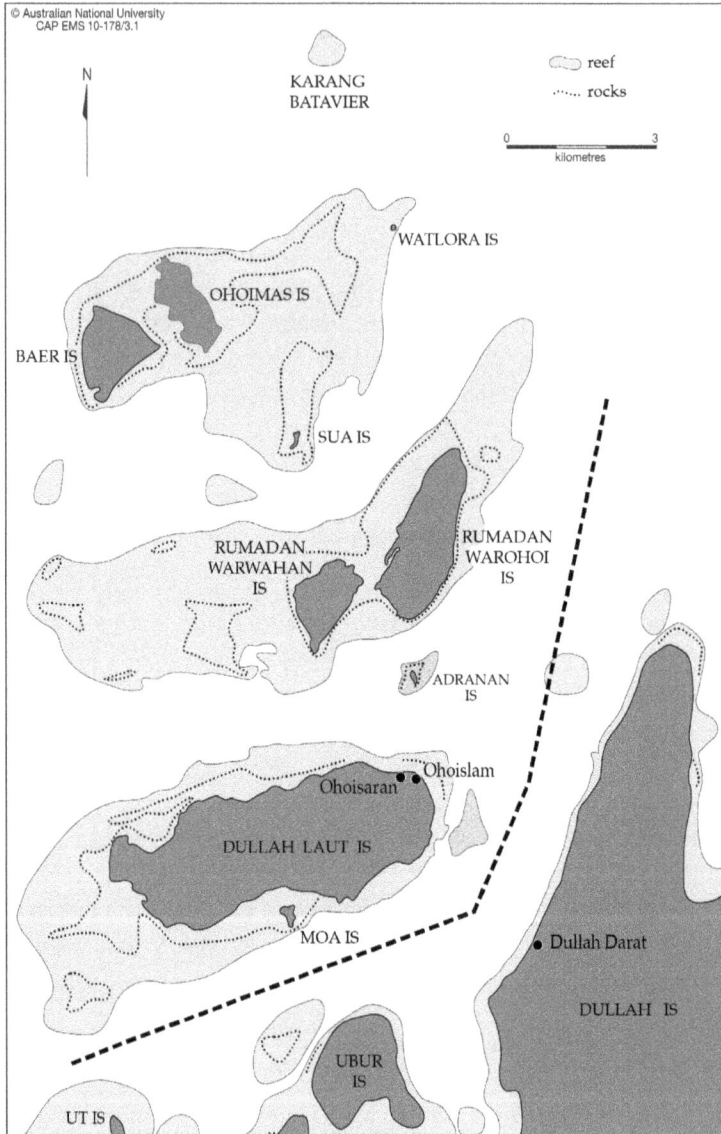

© Australian National University
CAP EMS 10-178/3.1

N

KARANG
BATAVIER

reef
rocks

0   3
kilometres

WATLORA IS

OHOIMAS IS

BAER IS

SUA IS

RUMADAN
WARWAHAN
IS

RUMADAN
WAROHOI
IS

ADRANAN
IS

Ohoislam
Ohoisaran

DULLAH LAUT IS

MOA IS

Dullah Darat

DULLAH IS

UBUR
IS

UT IS

Map 3-1: Dullah Laut territory.

Source: Author's fieldwork.

The sea surrounding these islands is relatively flat and covers a wide fringing reef. Even though there is no lagoon, the reef makes it possible for people to use simple and relatively cheap fishing technologies very effectively. Despite the simplicity of the technologies used, the fishing bounty of the area enables the Dullah Laut fishermen to be one of a limited number of major fish suppliers to the fish market in Tual. The competition between grouper fishing companies to gain access to the Dullah Laut waters is further evidence that these areas have excellent fishing potential.

Settlement Layout

The village settlements are located on the eastern coastal tip of Dullah Laut Island. Before motorised boats, people used to travel to Tual—the capital city of the Southeastern Maluku Regency—by paddling to the village of Dullah Darat on Dullah Island, three or four kilometres to the east, and then traveling by land for another 14 kilometres to the south. Since outboard engines have become available, travel has become much faster and easier. For those with their own motorised boats, they can go directly to Tual and even to more distant villages. Those who do not have their own boat can use the public passenger boats that have been operating for the last 12 years. Two locally owned boats with 25 horsepower outboard engines provide a regular service from the settlements to Tual and return each day. These public services usually depart from the settlements in the morning and reach Tual in around an hour. They return to the settlements around midday.[2]

There are two settlements in Dullah Laut. One is called Ohoislam (Muslim settlement) and the other is called Ohoisaran (Christian settlement). As their names imply, the Muslim settlement is exclusively populated by Muslims and Catholics populate the Christian settlement. It is worthy to note that segregation on the basis of religious beliefs is common in Maluku and even in most of eastern Indonesia rural villages.

Plate 3-1: Sea view of the Ohoislam settlement (2009).

Source: Author's photograph.

---

2   When I went back to the village in December 2009, some outboard boats were being used to transport people from Dullah village to Dullah Laut every 15 minutes or so. As a result, the boat services from the village directly to Tual ceased.

Plate 3-2: The under construction stone dock in Ohoislam (1996).

Source: Author's photograph.

Approaching the village from Dullah Darat or Tual, Ohoislam settlement comes in to view first (see Plate 3-1). Access to the settlement is relatively easy from any direction because the coast is protected from strong currents and waves. A prominent stone dock, which was under construction when I did my fieldwork and is located on the southeastern edge of the settlement, is notable for its socio-political relevance rather than the fact that it is the only dock on the island (see Plate 3-2). In particular, the dock was built by Mr A. Rahaded and his supporters who are considered to be in political opposition to the modern village head. The construction of the dock was one of the ways in which Mr A. Rahaded and his group expressed their rejection of the modern village head's power (detailed further in Chapter Six).

Before I describe the Muslim settlement further, let me briefly explain the social stratification of the Kei people (for detailed discussion see Chapter Four). This is important because this stratification is reflected in the housing pattern of the settlement and it also underlies political segregation in the village. Individuals in Kei are divided into three social strata: these are the *Mel* or *Mel-mel* (the nobles); the *Ren* or *Ren-ren* (the commoners); and the *Iri* or *Iri-iri* (the former slaves). These three strata form distinctive groups and inter-marriage is forbidden. Traditionally, the *iri* were owned by the *mel* while the *ren* were considered 'free people.' In contemporary life, these forms of relations are contested and political alliances between different strata can occur and change over time depending on the interests at stake and the context of the interaction. Being sensitive to the use of the term 'slave,' I will use the vernacular terms *mel, ren,* and *iri* for the rest of the book except in cases where the use of the terms 'slave' or 'former slave' are unavoidable.

The socio-political segregation of the Muslim settlement is best described by a bird's eye view of the settlement layout. With reference to Figure 3-1, the house on the beach marked 'X' is one of five on the block that faces a footpath that leads to the Islamic Elementary School at the back of the settlement. This footpath divides the settlement into two based on the social rank of its inhabitants. On the left hand side of the footpath is the block comprising solely of the *iri* family houses. On the right of the path are the houses belonging to *mel* families, except for three houses next to 'X' and three others at the back of the settlement. Secondly, the footpath demarcates the political followings in the settlement. Although Mr A. Rahaded's house (5) and some of his supporters live on the right side of the path, most of his support base is on the left side of the path and this is where political activities of this group are held. Also on the left of the path, a small mosque (3) has been erected. On the right side of the path is where the communal activities of the village head and his supporters are held and only this political group uses the big mosque located on the right of the path. The Islamic Elementary and Junior High schools (1 and 2) and the Community Health Centre (4) are not segregated and both social ranks and political groups use these public facilities.

About eight hundred metres to the east of the Muslim settlement, passing through the Muslim cemetery complex and coconut plantations, is the Christian settlement. Although it is approximately half the size of the Muslim settlement, the general layout of the Christian settlement has the same rectangular design but with a number of different characteristics.

To look in detail at the Christian settlement, the church is the most suitable point of reference (see Figure 3-2 and Plate 3-2). The church is located in the centre at the back of the settlement. It shares the same block as three slave family houses and a noble house. Across a foot path at the back of this block is the Catholic Elementary School. The church faces a field which is used by the youth to play soccer and is more importantly known by the metre high tower in its far left hand side corner.[3] This tower is the sign of the 'centre' (*woma*) of the settlement. In the Kei Islands, *woma* is a 'proof of origin' which means that the settlement has been authentically established by the island's ancestors. When people question whether a settlement is a place of origin, meaning the location where the ancestors started their communal lives, they will simply look for its centre. Unless a centre is found, the settlement will be considered to be newly established.

---

3   On my return visit to the village in December 2009, I found that this field had become an agricultural plot where people planted cassava, corn (maize), and some vegetables (see Plate 3-2).

Figure 3-1: Ohoislam settlement layout.

Source: Author's fieldwork.

Plate 3-3: The church and converted field in Ohoisaran (2009).

Source: Author's photograph.

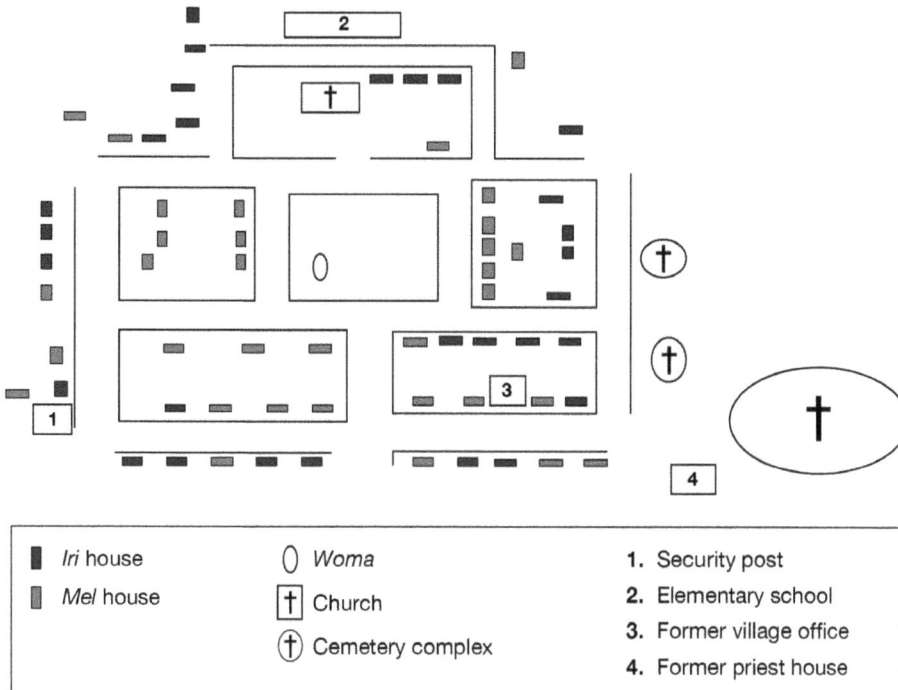

Figure 3-2: Ohoisaran settlement layout.

Source: Author's fieldwork.

In the Christian settlement, neither social rank nor political affiliation can be identified from the position of houses in a block. The *mel* and *iri* family houses are intermingled. Looking more closely, the *mel* houses surrounded by brick fences can be clearly distinguished from the wooden fenced houses of the *iri* families. According to some informants, differentiation on the basis of fence type was introduced in the early-1990s at a time of conflict between the *mel* and the *iri* involving the construction of the church.

Demography

According to a census I conducted in December 1996, the total population of Dullah Laut was 1 231 persons, of which 904 (73 per cent) lived in the Muslim settlement and 327 (27 per cent) in the Christian settlement. In terms of gender, there were slightly more males (622) than females (609) totaling a ratio of 1.02. Looking at each of the settlements, I found that the male to female ratio was 0.98 at the Muslim settlement and 1.14 at the Christian settlement. These calculations were based on the fact that at Ohoislam and Ohoisaran the males numbered 448 and 174 respectively, while the females numbered 456 and 153 respectively.

Calculating those who were born in Dullah Laut but who lived outside the village—ranging from the Kei Islands to as far as Jakarta—the gender ratio is higher than those who lived in the village. Of the 568 people who live outside Dullah Laut, 316 were males and 252 were females: a ratio of 1.25. Looking more closely at each settlement, comparatively more males moved away from the Muslim than from the Christian settlement. Of the 463 people who moved away from the Muslim settlement, 263 were males compared with only 53 males of the 105 people who moved out of the Christian settlement.

Table 3-1 shows the distribution of the population of Dullah Laut by age category. It is interesting to see that the distribution does not take the pyramid shape (see Figure 3-3) common in rural areas throughout Indonesia. In Dullah Laut, the elderly, represented by people of more than 60 years of age, was a larger group than that of middle ages categories (41-50 and 51-60 years old). This means that the life expectancy was quite higher. Nonetheless, we can see that the total productive age (13-60 years old) was higher than that of non-productive ages (0-12 and >60) at 655 and 565 people respectively.

Looking at each category, we can see that the largest proportion of the population falls into the category of 0–12 years old. This category constitutes 35 per cent of the population of which nearly half (17 per cent) are five years or under. The numbers of inhabitants decrease with increasing age, except for the category of more than 60 years old. The categories 13–30, 31–50 and over 50 years old comprise 25, 21 and 18 per cent of the population respectively.

Table 3-1: Population of Dullah Laut, December 1996.

Age	Ohoislam		Ohoisaran		Total
	Male	Female	Male	Female	
0–12	155	173	57	47	432
13–30	123	114	38	27	302
31–40	60	59	20	21	160
41–50	36	34	18	14	102
51–60	28	28	18	17	91
> 60	39	44	23	27	133
No data	7	4			11
Total	448	456	174	153	1231

Source: Fieldwork census, December 1996.

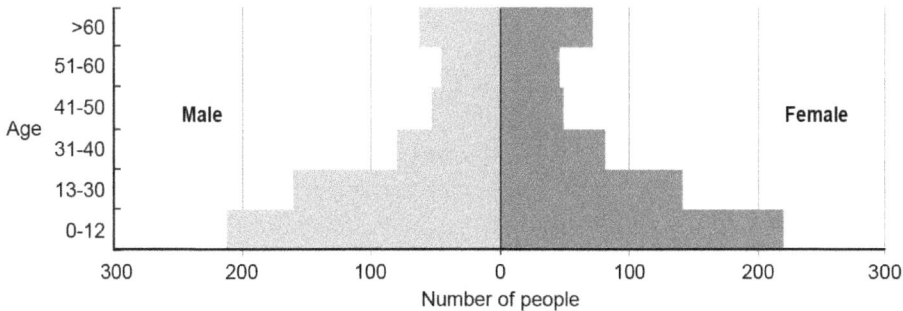

Figure 3-3: Age and gender distribution of Dullah Laut Village, 1996.

Source: Author's fieldwork.

There were 142 households in the Muslim settlement, living in 132 separate houses. This meant that some households shared their home with other families. There were in fact, some houses occupied by more than two households. At the Christian settlement, there were only 59 households in 58 houses. Only one house occupied more than one additional household.

In terms of occupations, the residents of Dullah Laut are mainly involved in farming and fishing. Almost all adults of both sexes do some farming activities. Thus, the question of occupation relates more to what they do other than farming. In that regard, fishing is the most popular occupation in the Muslim settlement with 113 people engaged in fishing activities. Assuming that the productive age is between 13–50 years old and considering that fishing is a male occupation, this means 88 per cent of the productive male population spends their working time at sea. In the Christian settlement, only 20 out of 78 males of productive age (26 per cent) earn cash from fishing. Furthermore, only a small portion of the population from both settlements work as civil servants, tailors, drivers, small-scale shopkeepers and teachers.

Kinship Groups

A *fam* is the most important kin group in Dullah Laut. As Geurtjens notes (quoted in Van Wouden 1968: 35), '*fam* is a male-centred kin group which recruits its members through affinity, blood relations, and "dependency"'. Affinity describes the situation in which once a marriage ceremony is held and the groom has paid the bridewealth (*wilin*), the bride becomes a member of the groom's *fam*. Blood relations refers to the fact that children born in a marriage in which bridewealth has been paid will be counted as members of their father's *fam*. Dependency relates to rank. Those who are from the lowest (slave) rank will be considered members of the *fam* on whom they depend. Their spouses and children will be included as well.

There are many different *fam* residing in Dullah Laut. I identified 94 different *fam* names: 66 names in the Muslim settlement, and 28 in the Christian settlement. Both settlements share nine *fam* names. However, the 94 *fam* names do not represent the real number of *fam* in the village. Some are only the names of individuals and these cannot be considered the names of kin groups. Some others are the names of groups who have only come to Dullah Laut recently and are relatively small in number. Therefore, I would suggest that the 'real' *fam*, in terms of kin group organisation and their significance in shaping Dullah Laut as it is today, number only 12 (see Table 3-2 for the names of these *fam*).

Table 3-2: The important *fam* and social stratification in Dullah Laut.

	Fam	Ohoislam		Ohoisaran		Total
		Mel	*Iri*	*Mel*	*Iri*	
1	Rahaded	37	34	7	–	78
2	Yamko	33	17	9	34	93
3	Henan	13	7	8	40	68
4	Rahawarin	17	13	7	1	38
5	Raharusun	51	59	32	80	222
6	Nuhuyanan	111	109	1	–	221
7	Songyanan	8	–	–	–	8
8	Fadirubun	13	54	–	–	67
9	Ngangun	11	15	–	–	26
10	Rumadan	9	20	–	–	29
11	Ohoimas	15	–	–	–	15
12	Mataraai	1	–	8	–	9
Total		319	328	72	155	874

Source: Fieldwork research, 1996.

The first seven names are the most important because they are considered to be the 'origin' (or original) *fam*. The other five have particular connection with the first seven or feature in the 'history' of Dullah Laut.

The first six *fam* and Mataraai are found in both settlements. The rest live exclusively in the Muslim settlement. All of these *fam*, except Songyanan, Mataraai, and Ohoimas, consist of both *mel* and *iri*. Songyanan, Mataraai and Ohoimas are exclusively *mel*. According to Table 3-2, the two largest *fam* in terms of membership are Raharusun and Nuhuyanan. In terms of social rank, it can be seen that membership of the *iri* is larger than that of the *mel* and the *iri* outnumber by more than double the number of *mel* in the Christian settlement.

A smaller kin group called *riin* exists in Dullah Laut. Literally, *riin* means room (of a house) and more specifically refers to a segment of a *fam*. Nevertheless, not every *fam* has *riin*. In fact, Nuhuyanan, Ohoimas, and Raharusun are the only *fam* that have *riin*. The *riin* of Nuhuyanan are Yahaw Warat, Vuur and Bal Ulab. The *riin* of Ohoi *mas fam* are Wahadat and Rasbal. The *riin* of Raharusun are Yayaan, Matwawan, and Watwarin. The *riin* of Nuhuyanan and Ohoimas *fam* are actually the names of brothers. The *riin* of the Raharusun are named after the relative age of three brothers: Yayaan is *riin* of the oldest; Matwawan of the middle; and Warwarin of the youngest. I was told that the segmentation of *fam* into *riin* usually occurred when brothers of a certain generation had an 'unusual' or 'remarkable' characteristic. Bal Ulab, for example, was famous for his physical and spiritual strength. This was why the ruler of Danar asked him to be his ally and appointed him a 'king' by sending *mas ayam vot* (gold medal in the form of crescent). Although Bal Ulab did not take the appointment to become a king, he was very well respected not only in Dullah Laut but also in the whole of Kei. This 'unusual' characteristic set him apart from his brothers Yahaw Warat and Vuur and resulted in the segmentation of Nuhuyanan *fam* into the three *riin*.

A *riin* as a distinct kin group is evident in Dullah Laut also. The Rahan Velav, a committee that makes certain decisions in the customary court, consists of *riin* Bal Ulab of Nuhuyanan *fam,* and Raharusun, Rahawarin, and Songyanan *fam*. This makes it clear that *riin* Bal Ulab of Nuhuyanan *fam* is considered to be a kin group on a parallel with Raharusun and Rahawarin *fam*. Another example is reflected in what is called the 'five houses' (*rahan en lim*) or the five keys (*kunci lima*). This group was created to promote cooperation between the five kin groups at the Christian settlement with obligations on all *fam* members to help each other on occasions of marriage, death, or other social occasions. As its name implies, this social group incorporates five kin groups: Raharusun Yayaan; Raharusun Matwawan; Raharusun Warwarin; Rahawarin; and Mataraai. Again, this indicates that Yayaan, Matwawan, and Warwarin as *riin* are considered to be kin groups in the same capacity as the Rahawarin and Mataraai *fam*. In the Muslim settlement, I once attended a gathering of Bal Ulab *riin*. The gathering was concerned with the expense of going on the pilgrimage to Mecca for an old member of the *riin*. This person, who was actually the leader of Nuhuyanan *fam,* started the meeting by recounting his problem. He had registered to go on the pilgrimage but unfortunately, did not have enough money to cover the cost. Another *riin* elder continued the speech by saying that this problem was not only a problem for the leader. It was a problem for the whole *riin* because if he failed to cover the cost—which meant failing to go on the pilgrimage—the standing of their *riin* would be degraded. It was decided at the meeting that all those attending should collect a certain amount of money within three days since the money had to be deposited on the fourth day. This meeting clearly demonstrated how key the *riin* kinship system is in developing cooperation between its members. When it was said that going on the pilgrimage was linked to the standing of the *riin*, it was clear that there was a sense of identity within the *riin*.

Both *fam* and *riin* are organised similarly. In theory, both are led by a leader called *yaman yaan*. As implied by the name—*yaman* (my father) and *yaan* (my big brother)—the leader of a *fam* or *riin* is usually the oldest male. However, in practice I often saw a *fam* represented by more than one elder or by different people, suggesting that the position of a *fam* or *riin* leader is not fixed. In the absence of a more eligible leader, another elder might occupy this position. Furthermore, a *fam* or *riin* leader is a person to whom people go for advice in difficult times. He is the one who is supposed to settle disputes before they are brought to a customary court in the settlement or village level. A *fam* or *riin* leader also represents his *fam* in dealings with other *fam* or larger social groups. The latter can be seen when attending a customary court. Every leader of an origin *fam* and/or *fam* associated with the issue being discussed in the customary court is reserved one or more seats. Every *fam* or riin leader may say something in the discussion before the court.

In return, a *fam* or *riin* leader is theoretically, one who deserves respect and obedience from his *fam* members and even members of other *fam*. In practice however, it is not necessarily so. For example, I sometimes observed younger people who did not agree with the ideas of their *fam* leader expressing their disagreement by grumbling in a back room of the house of the modern village head where important meetings were held.

Gender and seniority—in terms of both generation and age—is very important in the kinship organisation in Dullah Laut with the oldest males holding key positions. The position of *fam* leader which is usually held by the oldest man is an example of this. This position is also theoretically transferred to the oldest son. The practice of wealth inheritance is another example. Whenever parents die, the oldest son takes control of all the wealth of his family. It is in his power to determine whether the wealth will be distributed or not.

Marriage Patterns

Geurtjens found that *fam* were exogamous with women exchanged between *fam*. He further noted that the exchange pattern is not symmetrical. This means that two *fam* cannot exchange women directly, and that a *fam* which gives a woman (*mang ohoi*) to another *fam* (*yan ur*) will not receive a woman from them. There should be at least one other *fam* involved which provides a woman to the first. Barraud supports this finding. Using the term 'house' instead of *fam,* she suggests that the house is 'usually exogamous' (Barraud 1990b: 197). She notes that:

> the *yan ur* house is the groom's, while the *mang oho*[*i*] house is the bride's…. From the former wives are taken, while daughters and sisters are married to men in the latter. A man is not permitted to marry a woman from a house into which his sister has married. Such system is known as asymmetrical marriage…. (ibid.).

Geurtjens (quoted in Van Wouden 1968: 12) notes an exception to this pattern. He discovered that marriage between 'less prominent members' was somewhat different. While the practice amongst the nobles was matrilateral cross-cousin marriage (a man should marry his mother's brother's daughter), amongst the lower ranks a man married his father's brother's daughter. This results in *fam* endogamy. The existence of symmetrical exchange is also mentioned. He found that women were exchanged directly between two *fam* of the lower ranks (ibid.).

Interestingly, in Dullah Laut, marriage patterns are different again. There are indications that *fam* endogamy is relatively common among the *mel*. Of 214 marriages in the Muslim settlement, I found 39 (18 per cent) were endogamous. If we differentiate these cases by social rank, we can see that endogamy is more common among the *mel* than among the *iri*. There were 25 endogamous marriages amongst 122 noble couples. This means that 20 per cent of marriages among the *mel* were endogamous. By contrast, for the *iri* only 14 of 92 marriages were endogamous, only 15 per cent. The pattern is much the same if we take into account the 113 cases where one or both partners were deceased. There were 18 (16 per cent) endogamous marriages, most of which were among the *mel*. Of 69 deceased *mel* couples, 13 (19 per cent) marriages were endogamous compared with only five (11 per cent) among the 44 *iri* couples. Looking closely at the present couples, it emerged that Nuhuyanan was the *fam* in which endogamy was most common, with 18 (29 per cent) endogamous marriages among 63 Nuhuyanan couples. Furthermore, Nuhuyanan *mel* had 15 (31 per cent) endogamous marriages out of a total of 49 Nuhuyanan *mel* couples. Among the deceased couples, the highest rate occurred in the Raharusun *fam* with four (33 per cent) cases of endogamy out of 12 Raharusun couples. Nuhuyanan was in second position with five (26 per cent) cases out of 19 couples.

At the Christian settlement, the marriage pattern is similar to that of the Muslim settlement. Endogamous marriages occur with similar frequency. I noted that 15 out of 64 (23 per cent) marriages were endogamous. Dividing the cases by social rank, I discovered that three (14 per cent) out of 21 *mel* marriages were endogamous, while 12 (28 per cent) out of 43 *iri* marriages were endogamous. Among those who were deceased, I found more endogamous *mel* couples. Seven out of 27 *mel* marriages (26 per cent) were endogamous. In the *iri*, I found only five endogamous cases out of 41 marriages (12 per cent). These figures suggest that while endogamy has declined among the *mel* it has increased among the *iri*. Asked about the issue of endogamy, a *mel* informant whose marriage was endogamous told me that the church restricts marriage between close kin. He explained that this regulation had not been strictly applied until the last two decades. This may explain the reduction in endogamy among the *mel* but does not explain the pattern of increase among the *iri*.

The Muslim and the Christian settlement populations have different marriage patterns especially if we look at those who married outside their *fam*. In the Muslim settlement, many women are exchanged directly between two *fam*, while at the Christian settlement I did not see any cases which followed this pattern among either living or deceased couples. It is interesting that symmetrical exchange at the Muslim settlement is performed not only among the *iri* but also among the *mel*. We can see from Table 3-3 that of the seven origin *mel fam*, five are involved in symmetrical exchange. For example, we see that Nuhuyanan took 12 Raharusun women and in return Raharusun received seven women from Nuhuyanan. The Rahawarin *mel* who took three Nuhuyanan women gave four of their women to the *mel* of Nuhuyanan.

Table 3-3: Exchange of women among the origin *Mel Fam* in Ohoislam.

	A	B	C	D	E	F	G
A		1	–	–	1	–	1
B	2		–	–	–	1	6
C	1	–		–	2	2	–
D	2	–	1		–	–	4
E	1	2	–	–		–	12
F	–	–	–	–	–		–
G	2	–	–	3	7	–	

Note: Column=wife-giver; row=wife-taker; A=Rahaded; B=Yamko; C=Henan; D=Rahawarin; E=Raharusun; F=Songyanan, G=Nuhuyanan. Bold font indicates the existence of symmetric marriage exchange.

Source: Fieldwork research.

Among the lower rank, a similar pattern is evident. Table 3-4 shows the exchange between eight *iri fam*. For example, Rahaded took one and three women from Raharusun and Nuhuyanan respectively, and in return they gave three and one women to Raharusun and Nuhuyanan. Fadirubun took five Nuhuyanan women and gave two women in exchange.

Table 3-4: Exchange of Women among the _Iri Fam_ at Ohoislam.

	A	B	C	D	E	F	G	H
A		3	1	–	–	–	–	–
B	1		4	3	1	–	–	2
C	3	2		5	1	1	–	1
D	–	–	2		–	–	–	–
E	–	–	–	2		–	1	1
F	–	–	2	1	–		–	–
G	–	1	–	–	1	–		–
H	–	–	–	1	1	–	–	

Note: Column=wife giver; row=wife taker; A=Rahaded; B= Raharusun; C=Nuhuyanan; D=Fadirubun; E=Renleew; F=Renuat; G= Walerubun; H=Ngangun. Bold font indicates the existence of symmetric marriage exchange.

Source: Fieldwork research, 1996.

Inter-_Fam_ Relations

Marriage patterns are one of the instruments anthropologists employ to consider relationships between kin groups. From Van Wouden's (1968) _Types of Social Structure in Eastern Indonesia_ and Barraud's (1990b) _Wife-Givers as Ancestors and Ultimate Values in the Kei Islands_, it is obvious that there are hierarchical relationships between _fam_ allied by marriage. Wife-giving _fam_ (_mang ohoi_) are always considered to be superior to the wife-taking _fam_ (_yan ur_). Van Wouden (1968: 11, drawing from Geurtjens 1921: 303) puts it as follows: 'The _yan ur_ are required to acknowledge the primacy of the wife-giving _fam_ by paying them respect and offering them gifts at such family occasions as births, marriages, and deaths'. Barraud (1990b) even found that the Keiese put the wife-giving _fam_ alongside the ancestors in respect and importance. This is usually shown during life cycle ceremonies and religious rituals.

Evaluating inter-_fam_ relationships at Dullah Laut, I found a significant difference between Muslim and Christian practices. The death ceremonies I witnessed provide good examples of these differences. In the Muslim settlement, people brought a 'donation' (_yelim_) to the house of the dead person. The donated goods included consumables such as rice, sugar, coffee, tea, flour, and/or money. As well as being personal donations, these goods were also given as part of the _riin_ and/or _fam_ obligations. I observed that the origin _fam_ from both settlements brought their donations to the house of the dead, but I did not see any special 'donation' given to the dead family from their 'wife-taking' _fam_. The only times I heard the terms _mang ohoi_ and _yan ur_ used were during the speech made by the representatives of the dead person's family. By contrast, at the Christian settlement, in addition to consumables, customary goods were not only given but also exchanged. There was a 'customary seat' (*duduk adat*) where the wife-taking *fam* presented the customary wealth (*harta adat*) such as a gold bracelet which is called 'three tail gold' (*mas tail til*) referring to its quality. In the evening, the representatives of the *fam* of the dead person came to the mourning house bringing clothes for the dead. In return they were presented with a gold bracelet and an antique ceramic plate.

These examples are typical of the relations between *fam* in Dullah Laut in general. For the Muslims, the hierarchical arrangement between *fam*— specifically between wife-taker and wife-giver—did not appear to be an important character of inter-*fam* relationships, though this is not to say it has disappeared. An example of this was the political conflict led by a descendant of the traditional village head and the former modern village head and his son (the present modern village head). The conflict centered around which of the two was the rightful leader of the village as well as the distribution of money or materials the village received from a government subsidy provided by the central government. The traditional leader's descendant was the former modern village head's father's sister's son, meaning the traditional leader's descendant was the wife-taker of the former modern village head. For the Christians, hierarchical attitudes dominate and the wife-giving/wife-taker relationship prevails albeit with reluctance in some circumstances. This is why those who were dissatisfied and who considered the leadership of the current head of settlement too weak never expressed this explicitly. Another example was explained by Mr P. Rahaded, an informant in the Christian settlement who said that his support of the settlement leader was due to the fact that his wife was taken from there.

The Settlement Organisation

The organisation that oversees the kin group is the settlement (*ohoi kot* or also called *dusun*). It is a grouping of a number of *fam* who reside in a particular hamlet. As I mentioned earlier, Dullah Laut consists of two settlements and to some extent, each of these is governed independently by a Muslim or Christian settlement leader. However, although it has its own settlement leader, the practical leadership in the Muslim settlement is directly under the control of the village head. It is only at the Christian settlement that the settlement leader exercises some level of autonomy.

The settlement leader is responsible for keeping order in his settlement and handling all problems that arise. I observed that problems that were brought to a settlement leader mostly concerned conflict between members of more than one *fam*. This may be due to the fact that a *fam* leader usually makes an effort to solve problems that arise between members of his *fam*.

In dealing with complicated problems—such as conflicts that involve a large number of people or different social ranks—neither settlement leader[4] works alone. Each has a committee which consists of representatives of all origin *fam*, the *Ohoiroa Fauur*. When dealing with these problems, a settlement leader will call a customary meeting attended by all parties involved in the conflict and all representatives of the origin *fam*. Led by a settlement leader, the origin *fam* representatives will act as both judge and jury and will discuss the problem and decide on the appropriate solution.

The role of a settlement leader in a customary meeting depends on his personal power. A powerful leader will take on the role as overall leader and will be the key player throughout the meeting. This means that the final decision is in his hands although ideas from all committee members will be heard. A settlement leader who has less power will only act as a facilitator during the meeting. According to some elderly villagers—some of them *fam* leaders—the question of power was never an issue in the old days. The words of a settlement leader or other customary leaders were never disputed, at least not in their presence. Nowadays, it is said, people are very critical and arrogant. Regardless of where, when, and who, if they hear something they do not agree with they will argue. I would suggest this is because custom is not the sole source of power anymore. When I attended a customary court at the Christian settlement, I saw *fam* leaders (including civil servants and a retired policeman) openly challenging the ideas of the settlement leader. This forced the settlement leader to invite a military officer who'd participated in the meeting to bring the meeting to a close without any agreement reached. I will discuss this situation in more detail in Chapter Six.

The Village Organisation

In the course of its history, the political organisation of Dullah Laut has experienced several changes. It must be made clear that the people usually differentiate between two village political organisations. The first is the traditional village organisation which refers to the village as a traditional village. The second is the political organisation of the village under the Indonesian government regulation, particularly after the implementation of the *Village Government Law, No. 5, 1979*. In this regard, the village is called a modern

---

4   I consider the village head the settlement leader whenever he deals with internal problems in the Muslim settlement.

village. This section will discuss the traditional and modern forms of village political organisation in Dullah Laut and look at how villagers adapt to changing political circumstances.

As a traditional village, Dullah Laut is led by a single headman under the title 'wealthy person' (*orang kaya*). According to tradition, this position is handed down from father to eldest son. These positions could only be held by a noble (*mel*) member and they were always associated with the local history of the village. However, the first traditional village head in Dullah Laut was believed to be appointed by the Dutch in the early part of the twentieth century. It was after the second traditional village head that the position was transferred following the traditional regulations.

The duty of a traditional leader was to keep order in the village. I noted earlier that in times of conflict, it was the duty of a village leader to restore order if a settlement leader could not resolve an issue in his settlement. A traditional village leader also represented his village to the outside world. Thus, when the Dutch government wanted to make a connection with villagers, it was through the traditional leader that the Dutch officers would do so. It was also through the traditional village leader that a villager could make a connection with the outside world. If a villager wanted to travel out and needed a recommendation letter for that, it was the traditional village leader who would prepare the letter.

A traditional village leader did not work alone in managing his people and territory. Geurtjens (1921, quoted in Van Wouden 1968: 36–7) noted that several functionaries worked together with the village head. Three important figures were: 'lord of the land' (*tuan tan*); the 'attendant of the local spirit' (*mitu duan*); and the 'Islamic religious official' (*lebay*) to which I would add head of settlement (*bapak soa*). 'Lord of the land' was the official lord of all village lands and his role was crucial in allocating land territory. He was the one people went to whenever they wished to make a new garden and he knew most about land distribution between people in the village and the boundaries between neighbouring villages. Therefore, he played an important role in solving land and sea ownership disputes. Both the attendant of the local spirit and the Islamic religious official were responsible for dealing with affairs relating to the ancestors, local guardian spirits, and God. As I mentioned in the preceding section, the head of a settlement was responsible for keeping order and resolving issues in his settlement.

However, two of these functionaries did not play an important role in Dullah Laut. The 'lord of the land' in fact was only mentioned in the narrative of origin. Moreover, none of my informants could recall the existence of the 'attendant of the local spirit'. This might relate to the fact that almost no rituals pertaining to indigenous belief have been maintained. Although the Muslim imam and the

Catholic priest might be seen as having replaced the former *lebay* position, their roles are quite different from the traditional role. They were never present at any customary court sessions during my research. It seemed that among these functionaries, only the head of settlement played an important role in helping the village head.

A traditional village head also worked with the traditional village assembly. In fact, it seemed that the village assembly was the ultimate power holder in the village. All of the traditional leader's decisions concerning important issues— such as those relating to territory—were to be made in consultation with the village assembly. The process for dealing with important issues was that the traditional village head should call a customary meeting attended by all members of the village assembly so issues could be discussed and resolution reached.

In Dullah Laut, the village assembly is a committee of origin *fam* called *Ohoiroa Fauur*. The members of this committee are the leaders of Henan, Rahaded, Yamko, Raharusun, Rahawarin, Nuhuyanan and Songyanan *fam*. Since the village is divided into two settlements because of religious differences, the representatives of origin *fam* are also taken from both settlements. Thus, each origin *fam* is represented by two leaders, one from each settlement. As a result of the representation from both settlements, the village assembly in Dullah Laut has a total of 14 members.

The Indonesian government replaced the Dutch when the Kei Islands were declared to be part of the Republic of Indonesia. Although, in the beginning, tradition still played a role in governing the village, new arrangements were introduced when the traditional village of Dullah Laut was split into two modern villages. These were Dullah Laut Islam and Dullah Laut Roma Katolik. This meant that the traditional Christian settlement which used to be under the coordination of a traditional village head located at the Muslim settlement became an independent village. Of course, this separation changed the nature of the relationship between the two villages significantly.

The *Law of Village Government (Pemerintahan Desa), No. 5, 1979*, intensified the presence of the Indonesian government in matters of village governance. Although, Dullah Laut was merged into a single village again, the law put aside the position of tradition almost completely introducing a new procedure for appointing the village head. Under the law, a village leader could be any one of the villagers who principally believed in one God, was loyal to the Indonesian government, healthy, aged 25–60, and had acquired at least senior high school education (Marsono 1980). The new structure of village functionaries (*perangkat desa*) was also different from that of the traditional one. The new village functionaries consisted of a village secretary (*sekretaris desa*), some program coordinators (*kepala urusan*) and settlements heads (*kepala dusun*). Additionally, the application of the village law also changed the name of the traditional village assembly (*saniri negeri*) to village deliberation council (*lembaga musyawarah desa*). The law also regulated the appointment of village functionaries and the village deliberation council which was different from the traditional regulation. Just as in the new process for the appointment of village head, all of these positions were no longer hereditary.

Interestingly, the people of Dullah Laut—and probably most of the Kei people— considered these institutional changes to be a supplementary arrangement, in the context that the institution of the modern village head was an addition to the pre-existing traditional village head arrangement. This assumption becomes clear when investigating the genealogical connections between the current and the former modern village heads. The preference for choosing a candidate who was the son of the former modern village head was still apparent even after the village law had been implemented. This is also clear if we examine the conflict concerning the village leadership, which has been taking place since the 1970s and continued at least until I finished my fieldwork in 1997. Along with using the Indonesian regulations and laws, the conflicting parties use tradition as the basic reference point for their arguments. Mr A. Rahaded, for example, demanded that Mr M. Nuhuyanan (former modern village head and father of the current village head) hand over his position as the village leader on the basis that Mr A. Rahaded was the descendant of three former traditional village leaders. These examples show that the structural changes from the traditional to the modern village structure were in name only. Those who were appointed to the new positions were those who held the position in the traditional village. So, members of the village deliberation council were leaders of origin *fam*.

Conclusion

This chapter highlights some characteristics of Dullah Laut Village and its people. An important feature worth mentioning here is that of social division. First, by tradition the inhabitants of Dullah Laut Village were divided into different social ranks—the *mel*, the *ren*, and the *iri*. Second, when Islam and Catholicism were brought to the village, they divided the population further based on religion. Finally, when a new political structure was introduced as the Kei Islands became part of the Republic of Indonesia, new notions of community and its social organisation were introduced. The application of the concept of modern village and its organisation brought about new divisions in the community. Desa Dullah Laut Islam, Dullah Laut Roma Katolik, *kepala desa* (head of the modern village), *dusun* (the settlement or grouping of a number of *fam* who reside in a particular hamlet), *anak desa* ('child village') and *perangkat desa* (village functionaries or officials) were added to the traditional vocabularies to join *negeri* (traditional village), *orang kaya* ('wealthy person', the title for a traditional village head), and *saniri* (committee of leaders in a village).

Interestingly, these social divisions were expressed—or at least marked—in the geographical layout of the settlements. In terms of social rank, the houses of the *iri* were located separately or fenced differently from that of the *mel*. In terms of religious division, the Muslims and Catholics lived in separate settlements in Ohoislam and Ohoisaran respectively.

In times of conflict, these divisions complicated the problem. The longstanding political conflict between a descendant of the village traditional leader, the modern village head, and the leader of the Christian settlement was not only reflected in the geographical layout of the village but was also instrumental in reinforcing existing divisions. It was apparent that those supporting the modern village leader were mostly the *mel* of the Muslim settlement. The descendent of the traditional leader's political followers were mostly the *iri* of the Muslim settlement, and the Christian settlement leader's followers were the Catholics at the Christian settlement. These political factions were also evident in the use of places of worship. The modern village's followers had prayers in the mosque while the supporters of the descendants of the traditional leader used their own small Islamic praying place (*musholla*). The Christian settlement leader's faction carried out religious rituals in the only church available in the village. As such, the political conflict in the village was interwoven with issues of social rank and religious sentiment.

This chapter also discussed kinship groups and marriage patterns. These social linkages may have had the potential to unite a divided community, especially given those who were not tied by genealogy might be connected by exogamous marriage alliances. However, it seems that marriage patterns have changed and the asymmetric marriage patterns—which were considered effective in creating wide-ranging alliances—have been replaced by symmetric and endogamous marriage patterns. This changing pattern might even have contributed to a worsening of the conflict.

4. Narrative of Origin: Social Organisation, Leadership and Territory

The oral history that describes the formation of a particular domain is referred to as *toom*.[1] Kei people believe that the events mentioned in a *toom* actually took place in the past. These events are important because they explain the process of creation of their social world. For the Kei people, *toom* is not only a history of their origin but also the most important source of traditional claim over positions and objects. Reference to a particular narrative of origin is required to legitimate any claim, thus for the people of the Kei Islands, the narrative of origin is the foundation of their tradition.

By explaining some Kei narrative of origins, the following discussion is aimed at showing that as the foundation of tradition, *toom* has more than one version and each is subject to multiple interpretations. As the basis of claims over positions and objects, these characteristics provide the basis for contestation. While people might argue that this is the sign of the flexibility of tradition, I would suggest that this can create problems because the nature of the *toom* makes it possible for people to craft, modify or even develop it for their own interests. When two or more people or groups with opposing interests are involved in such activities, conflict is unavoidable.

Narrative of Origin

The issue of origin is important for most Austronesian-speaking societies. In his article 'Origin Structures and Systems of Precedence in the Comparative Study of Austronesian Societies', Fox (1995: 34) states that 'among the Austronesians, the concern with origins represents a vital orientation, a basic epistemological stance, toward persons and objects in the world' (see also Fox 1996). Kei islanders are no exception to this. Knowing one's origin is not only a matter of understanding 'history,' but it is also a matter of justifying one's position in relation to others: who is ruling and who is ruled. In relation to ownership, it is a matter of who controls what, when, and where. In the Kei Islands, the discussion of origin is indeed a discourse of precedence.

---

1   Kaartinen 2009(a) spells it as *tum*. I choose *toom* after the double 'o' sound pronunciation of the word by the Kei Kecil islanders.

In the Kei Islands, the crucial characteristic of the origin narrative also focuses on issues of 'installing the outsider inside' (Fox 2008). Although the journeys of particular persons are mentioned, the most important part of the narrative of origin is the meeting of particular persons with the former inhabitants of the domain. This meeting is crucial because this is when negotiations concerning the distribution of power take place which then determine position in the domain. This process illustrates that the origin structure of the Kei people is more concerned with the creation of a domain rather than tracing the story of the ancestors' 'path and the road' as the Atoni do in Timor. (Fox 1988:12).

Alternatively, Kei Island narratives of origin can be differentiated into three categories on the basis of the social status of the players. The first category consists of narratives that only concern the establishment of a domain by the indigeous (cf. Kaartinen 2009(b); 2010). The second category recounts the meeting of immigrants with the native inhabitants of the Kei Islands. Narratives of this category become the basis of conflicts over precedence between different social ranks, particularly the *mel* and the *ren*. The third category includes all narratives that describe the meeting of different groups of immigrants. This type becomes the main source of the legitimisation of precedence among different *fam* of the same social rank, particularly the *mel*. The main issues of contestation within and between social strata mentioned in the narratives are issues of 'government' or domain leadership and issues of controlling the domain's territory.

This chapter provides examples of the second and third categories of narrative and shows how these narratives are used to explain the existence of social stratification and the distribution of power within and between different social strata. The first example is one recognised by most Kei people, while the second example is a specific narrative taken from Dullah Laut Village. The details of the narrative are unique to the Dullah Laut people. However, the theme—the meeting of different groups of immigrants—is more common. I believe most, if not all, traditional domains in the Kei Islands have similar versions.

Social Stratification

As briefly mentioned in Chapter Three, the people of Kei are divided into three social strata: the 'noble' (*mel* or *mel-mel*); the 'commoners' or 'free people' (*ren* or *ren-ren*); and the 'former slave' (*iri* or *iri-iri*). Laksono (1990) believes that the original structure of the three categories was not a vertical relationship with the *mel* at the top, the *ren* in the middle, and the *iri* at the bottom. He argues that the original differences between the *mel*, *ren* and the *iri* were based on whether they were indigenous or immigrant. Regarding the relationship between the *mel* and the *ren*, based on historical fact he suggests that:

the *ren-ren* were neither under nor above the *mel-mel*; both basically agreed that they were supposed to live together in a relationship of equality in which the *ren-ren* held the office of *teran nuhu* [or *tuan tan*, meaning lord of the island] and *mel-mel* held administrative office (Laksono 1990: 110).

It was only between the *mel* and *iri*—both considered immigrant groups—that relationships formed a hierarchical structure, whereby the *iri* was inferior in relation to the *mel*.

Furthermore, Laksono found that the current disagreement between the *mel* and *ren* was due to the introduction of a new hierarchical order by the Dutch. During their occupation, the Dutch granted certain administrative territorial titles such as *raja*, *orang kaya*, *kapiten* and *majoor* to their local collaborators.[2] They also issued letters of appointment and distributed knobbed canes (*rottingknoppen*) as a sign of the appointment. Since the appointment not only granted administrative rights but also territorial authority, this meant that the *mel* became the dominant group holding power. On the other hand, the *ren*, who were neglected by the Dutch, lost their territorial power as well as their balance of power to the former administrative authority of the *mel*. As a result, their position slipped to mid-rank between the *mel* and the *iri*.

For the most part, I agree with the above reading which differentiates between both the *mel* and *ren* and the *mel* and the *iri*. However, after reading the narratives of origin, I concluded that the *mel* not only adopted the new Dutch-introduced hierarchical structure, but also took a more active role of transforming their narrative of origin to imply superiority over the *ren*. Based on the *mel's* version of the narrative of origin, the two groups can claim to have never been equal in rank. This asymetric relationship has been the defining feature of their original relationship with the *ren*.

To make my point clear, I will discuss the Kei Islanders' narratives of origin. The people believe that the native inhabitants of the Kei Islands sprang forth from the earth or sea having emerged from animals and plants. These people were considered the first inhabitants of the islands and owners of the land and sea. As a consequence they were entitled to hold the title of 'lord of the land' (*tuan tan*) referred to as 'free people' (*ren*).[3]

---

2   As was common in other parts of Indonesia, the Dutch used the policy of indirect rule. This meant that at a local level, the Dutch did not create new political structures but used the existing political structures for their political and economic interests. In the Kei Islands, since the local political leadership was traditionally in the hands of the *mel*, it was the *mel* who were appointed to be the collaborators with the Dutch. It seemed that the Dutch were not aware that local tradition distinguished between political and territorial leadership and that these two issues were contested between and within different social ranks.

3   Actually, there is a narrative that asserts that some native inhabitants were considered to be the *mel*, however the *mel* consider this to be an exception.

Once upon a time, immigrants from various places[4] came to Kei and met the lord of the land. For various reasons these immigrants were incorporated into Kei society. Renyaan (1990: 3) in 'The History of Kei Tradition' notes that various versions of these narratives considered the immigrants to be smart, brave, and rich. These characteristics led them to win various physical tests and contests of spirituality against the native inhabitants. Some versions even mention that due to these superior traits, the native inhabitants invited the immigrants to live together with them and surrendered their territory and lives to be governed by the immigrants. To cite an example, here is an excerpt of the narrative from Englarang, Kei Besar, written in 1959 by Ahmad Rahawarin, the traditional village leader of Englarang:

> Balaha Rahawatin was the ruler controlling the territory of Englarang/ Ubohoifak, its sea and land and Ren-ren Hoerngutru Yelmesikrau. He was appointed by a leader of Ren-ren Hoerngutru Englarang and given the name of Lord of the Land of Englarang (Hemar). Thus, the lord of the land admitted that Balaha Rahawarin became their lord and leader, controlling all their possessions and Ren-ren Hoearngutru was ruled by Balaha Rahawarin for ever; for generations to come, Balaha Rahwarin was obliged to support Ren-ren Hoerngutru and Ren-ren Fuartel in time of need according to their custom (Adhuri translation).

Another version which is typically supported by the *ren* indicated that the installation was based on a mutual agreement on the distribution of rights between the two parties. The native inhabitants continued to hold their power over territory while the immigrants were given the right to rule the domain. Both versions still attest that the native inhabitants held the title 'lord of the land,' but the former version makes it an official title without any real control of their territory. In this version, 'lord of the land' is only understood as 'those who know the territory' while the latter version acknowledged the right of the lord of the land to control all issues pertaining to the territory of their domain.

Looking at the first version of the narrative, it is obvious that the immigrant *mel* asserted their superior position over the indigenous *ren* from their very first meeting. Even before they negotiated the distribution of rights, the immigrants were ascribed superior traits. I believe that this sense of superiority is why they put strong emphasis on issues of social boundaries in what they called the 'Law of Red Blood and Spear from Bali' (*Hukum Larvul Ngabal*). The following is a condensed version of the background narrative of the declaration of *Hukum Larvul Ngabal*:[5]

---

4 These immigrants were known by their place of origin such as from Bali and Sumba (*Mel Bal Sumbau*), from Luang and Maubes Islands (*Mel Luang Maubes*), and from Jailolo and Ternate (*Mel Delo-Ternat*).
5 This version is shared between the *mel* and the *ren*, but the interpretation was based on the *mel* concept.

The narrative begins with Kasdew and Jangra, two persons from Bali. After being installed within by the acceptance of the natives, Kasdew and Jangra, who then held the title of the great (*hilaay*), each developed their own domain in Ohoivuur (the present Letvuan village) on Kei Kecil Island, and Ler Ohoilim on Kei Besar Island. The social situation in Kei was in disorder at that time with crime, incest and other immoral acts occurring on a daily basis. Stimulated by these circumstances, Tebtut, the son of Kasdew, attempted to unite some *hilaay* around his domain. He held a meeting with *hilaay* from nine domains. The meeting declared a law called the Law of Red Blood (*Hukum Larvul*). This name derives from *lar* (blood) and *vul* (red) which was the blood of a buffalo slaughtered during the meeting.[6] The blood was a sign of the oath spoken by the nine *hilaay* that they had come to an agreement to uphold the *Hukum Larvul*. These nine *hilaay* were the origin members of the nine groups (*lor siwa*) (Adhuri translation).

A similar scenario was arranged in Kei Besar. Jangra held a meeting attended by the five heads of the hamlet, or *hilaay*. This meeting declared a set of laws called the *Hukum Ngabal*. The name *ngabal* refers to the spear (*nga*) brought by Jangra from Bali (*bal*). On this occasion Jangra slaughtered a whale (*lor*) and distributed it to the *hilaay* from Fer, Nerong, Uwat, Tutrean and Raharin, who got the head, stomach, tail, fin and teeth respectively. These five *hilaay* were core members of the five group (*lor lim*).

These laws were disseminated to the whole archipelago at the same time that the two groups were recruiting new members. Every new ally was appointed as a king and given a certain token reflecting their acceptance either as five or nine group members and applying either *Hukum Larvul* or *Ngabal*. Several wars broke out between the two groups as a result of their competition before they finally came to a peaceable agreement which united the *Hukum Larvul* and *Ngabal*. Ever since, the kingdoms of both sides have erected *Hukum Larvul Ngabal* as a single entity of their 'basic law'.

Returning to the issue of social boundaries, one could look at the contents of *Hukum Larvul Ngabal*. The law consists of seven[7] points, namely:

---

6   The buffalo mentioned in this narrative might not be the animal we call buffalo now because it is not native to the Kei Islands.

7   Some informants believe the law only has five points saying that the first two, the fifth, and the sixth are each a single verse. The King of Watlar told me that those who felt the *Hukum Larwul Ngabal* consisted of five verses were those who wanted to associate the law with the Pancasila (the Indonesian five pillars). However the different versions do not affect the content of the law.

Hukum Larvul Ngabal			Law of Red Blood and Spear from Bali
Hukum Nevnev	1.	Uud entauk na atvunad	Our head rests on the nape of our neck
	2.	Lelad ain fo mahiling	Our neck is respected, glorified
	3.	Uil nit enwil rumud	The skin made of soil covers our body
	4.	Lar nakmot na rumud	Blood is contained in our body
Hukum Hanilit	5.	Rek fo kilmutun	Marriage should be conducted properly so it can be kept in its purity
	6.	Morjain fo mahilin	The woman's place is respected, glorified
Hawear Balwirin	7.	Hira ini fo ini, it did fo it did	Theirs is theirs, ours is ours

Locals recognise three categories within the *Hukum Larvul Ngabal*. The first four points are considered to share the same theme concerning the principles of general conduct and are called *Hukum Nevnev*. Interestingly, using the upper part of the human body as an analogy, the first issue raised is the 'head', the focal point by which all parts of the human body are controlled. The most important concept deriving from this point is the unquestioned obligation to obey and glorify the ruler. In religious terms, this must be an obligation to worship god (*duad*). Regarding the social structure, the *mel* is the 'head' to which the *ren* and the *iri* are obliged to offer their submission. The second, third, and fourth points sustain the first and describe the specific obligation to respect life (point 2), not to gossip about others' misbehaviour (point 3), and not to attack others (point 4).

Points five and six—called *Hukum Hanilit*—concern issues related to women and marriage. The crucial topic here is the question of 'who may marry who'. The only answer to this question is rank endogamy: *mel* should only marry *mel*; *ren* with *ren*; and *iri* with their own kind. This is what the term 'purity' in point five refers to. Marriage outside this arrangement is subject to punishment. The most severe punishment—exclusion—occurs when a lower rank male marries an upper-rank female. The issue of sexual misbehaviour, which is the main concern of the sixth point, is also subject to the boundaries of social rank. Punishment for sexual harassment (impregnation, touching a woman's body or other types of harassment) within a single rank is always negotiable. But, for example, if an *iri* male harassed a *mel* woman he is subject to severe punishment. By contrast, sexual harassment of a male noble towards a lower-ranked woman is not subject to punishment, but is covered up. This different treatment shows that *Hukum Nevnev* is most concerned with maintaining boundaries rather than defining universally applied proper behaviour.

The last point—called *Hawear Balwirin*—regulates ownership, a very special issue. Interestingly, the focus of this point is not only material goods such as land, houses and clothes, but also social boundaries. Ohoitimur (1983: 64) notes the complete version of this point as:

Hukum Larvul Ngabal

| Hawear Balwirin | 7. | Hira ni ntub fo i ni, it did ntub fo it did, mel fo mel, ren fo ren, iri fo iri, teen fo teen, yanyanat fo yanyanat, yaan fo yaan, warin fo warin | Theirs is theirs, ours is ours, the mel is the mel, the ren is the ren, the iri is the iri, the parent is the parent, the child is the child, the oldest is the oldest, the youngest is the youngest |

The point to make about the *Hukum Larvul Ngabal* is that after declaring their superior position as rulers, the *mel* drew a boundary distinguishing themselves as the 'head' from the *ren* (and the *iri*) who were obliged to pay homage (*Hukum Nevnev*). This boundary was made clear by the prohibition of inter-marriage (*Hukum Hanilit*). Losing the chance to be linked by a marriage alliance meant losing one of the ways in which contestations of precedence could occur.[8] Finally, even in *Hawear Balwirin* social boundaries are stressed. The *mel* is *mel*, *ren* is *ren* and *iri* is *iri*. If their position was not assured as superior, I don't believe the *mel* would have made the boundaries so firm.

I would now like to turn specifically to the relationship between the *mel* and the *iri*. Unlike the *mel* and the *ren*, there is no question that the *mel–iri* relationship was hierarchical. There were no circumstance that could reverse their relationship—the *mel* was always superior to the *iri*. To be exact, their traditional relationship was that of master and slave. The noble was the lord or master, while the *iri* was the slave.

In analysing the original structure of the relationship between these two parties, Reid (1983) distinguishes between 'closed' and 'open' systems of slavery. He notes that 'a closed system of slavery may be defined as one oriented primarily towards retaining the labour of slaves by reinforcing their distinctiveness from the dominant population' (ibid.: 156). By contrast, 'open' slavery is defined as 'acquiring labour through the capture or purchase of slaves, and gradually assimilating them into the dominant group' (ibid.: 158). Using this division, slavery in the Kei Islands clearly took the form of a closed system. Once a person became a slave, there was no chance for him or her—or even their descendants—to return to his or her former status. Geurtjens observed other distinctive features of the slave. In his article 'De Slavernij of the Kei-eilanden', published in *De Java-Post*, 19 May 1911, Geurtjens noted that the slaves took care of almost all of their master's work and that their dress was regulated. The slave was not allowed to wear colourful clothing and had to wear a sarong above the knee. The word *sien*, meaning bad or ugly, was also added to their name. Geurtjens also mentioned that slaves were generally degraded, even in relation to god.

---

8 An example of how marriage strategies are used to achieve and maintain precedence can be found in Fox (1994).

One of the reasons why slaves were considered so low was because of the narrative of their path into bondage. According to local narrative, there are two main ways a person becomes enslaved: either captured during war or through judicial punishment. In ancient times, serious crime such as murder[9] or incest was punished by death, which usually involved sinking the culprit into the sea. Before the execution, however, the guilty person would be 'auctioned'. If someone bought him or her by paying certain customary wealth—such as an antique canon, gong or gold—the punishment would be cancelled and the wrong doer became the slave of the purchaser. The purchaser had to be wealthy because the price was high. In relation to both capture and purchase, the slave was considered to be polluted or in terms of rights, dead. This is why intermarriage was prohibited, enabling the noble to maintain their 'pure' blood. Since the slaves had no rights and only obligations to their master, it was logical for all aspects of slave life to be controlled by the master.

The economic and social benefits slaves provided their masters—such as free labour and social standing—ensured masters provided some support for their slaves, albeit of a low standard (see Reid 1983 for some examples). In this regard, Geurtjens (1911) observed that the nobles provided their slaves with basic needs and saw to ceremonies such as marriage and death. It was also the noble's responsibility to punish slaves for any wrongdoing.

Before looking at the contemporary slave and master relationship, I will discuss how these relationships were handled in the past starting with the story of Beruntung, a little Papuan boy who was captured in the Papuan War. When the Papuans were defeated in the war, he was the only person left after the battle of Ohoimas Island (Map 3-1). All the others were either killed or ran away. Masen father (*yaman*), the war commander of Dullah Laut from Rahawarin *fam*, brought the boy back to Dullah Laut as a token of their triumph. From that time on, the boy was named Beruntung and considered to be the possession of Masen's father. Some time later, when one of the Rahawarin *mel* members intended to go on the pilgrimage to Mecca, they sold Beruntung to the Rahaded *mel* for cash. This then transferred Beruntung from Rahawarin into Rahaded hands.

Another example was provided by a *mel* member of Yamko *fam*. I was told that his *fam* had obtained an *iri* member as a marriage gift (*lov fen-fen*) from their wife-giver (*mang ohoi*). What he meant by *lov fen-fen* was the gift of an *iri* given to *mel* members when members of Yamko *mel* married women from another *fam*.

---

9   Some killings, particularly those concerning a woman's dignity or territorial defence, were considered to be justified and those who killed for these reasons were not punished.

In this context, the *iri* was considered to be part of the 'accompanying goods' (*barang bawaan*) of the bride. Theoretically, the *iri* was supposed to help the bride to fulfil her duties as the wife of a man from another *fam*.

Some elderly people described to me how the *iri* were also under the control of the *mel* economically. I was told that the *iri* were a source of free labour. They had no choice when their master demanded labour, either for household duties—such as cleaning and collecting firewood and water—or for agricultural work such as opening or clearing gardens. As a result, it was very common for the *mel* who owned many *iri* to have many large gardens.

Politically, the *mel* controlled the *iri* as well. The *iri* had no right to be involved in any political decision-making processes, and political discourse and practice in the village was under the control of the *mel*. In Dullah Laut, for example, all political decisions were made at meetings of all origin *fam* representatives, village functionaries such as the traditional settlement head, religious leaders and the head of the village. People holding these various positions were exclusively the *mel*.

The present relationship between the *mel* and *iri* is quite different. Despite the fact that they are still called *iri* and looked down on, slavery is now a thing of the past. Referring to Reid's classification (1983), the relationship between the two might now be seen as a transitional system. On one hand, it is not a closed system any more because the *iri* enjoy some degree of freedom. But on the other hand, it would be difficult to say that their relationship has become a totally open system because to some degree, they are still considered a distinct social group, the boundary of which is kept through strict prohibition of inter-class marriage. The contemporary relationship between the *mel* and *iri* in the villages of Dullah Laut in the Kei Kecil Islands and villages on Kei Besar Island illustrate this situation clearly.

Unlike the picture of social stratification for Kei Islanders as a whole, other social orders have evolved in local areas including the creation of villages that are populated by one social rank only. Barraud (1990b: 196) for example, observed that the population of Tanimbar Village on Kei Tanimbar Island were all *mel*. In contrast, on Kei Besar Island, the population of Sather Village is composed of *ren* and the populations of Ngan, Rerean, Watkidat, Ohoilean and Uat hamlets are entirely *iri*.

The reasons for this diversity in social organisation lie in the formation 'history' of these villages and hamlets. The King of Tabab Yamlim on Kei Besar Island once explained to me the history of Ngan, Rerean, Watkidat, Ohoilean and Uat hamlets. He said that when his grandfather was king in the mid-nineteenth century and during his own time in 1935, some *iri* families were moved out of villages under their control. During these times, the issues behind the forced resettlements included population pressures and preventing intermarriage between people from different social ranks to maintain purity of the upper class. In his grandfather's time, there was also movement of *iri* families as punishment for sexual harassment of a *mel* woman. These people were ordered to establish new hamlets at Ngan, Reran, Watkidat, Ohoilean and Uat. Since marriage between social ranks is prohibited, these hamlets remain exclusively populated by *iri*. Some informants in Dullah Laut told me that the situation in their village was implied in the narrative of origin. When the immigrants came to the island, Landlord Henan—who was most probably a commoner—drove the immigrants inland and they became the 'disappearing people' (*orang ilang-ilang*). Or alternatively, as the narrative describes, they vanished because of a lack of members leaving only the *mel* and the *iri* on the island.

During my fieldwork in Dullah Laut, I observed that the current relationship between the *mel* and the *iri* was not the same as the historical accounts provided by Geurtjens or my own informants. Economically, the *iri* are now independent. I did not see any *iri* who worked in the house of their master. If the nobles needed labour for their garden, they might ask their *iri* to work for them, but they would pay them for it. Thus, the size of a mixed crop garden owned by a family no longer reflects the number of *iri* they have. Now, it depends on the size of the family, their willingness to work, and their capacity to pay people. This situation has made it possible for some *iri* to become wealthier than the *mel*. In fact, some *iri* families could afford to send two of their members on the pilgrimage to Mecca. I even found one *iri* family who'd supported their master's pilgrimage to Mecca. I was told that this *iri* family had contributed four million *rupiah* (approximately US$1739) which amounted to more than half of the total cost of around seven million rupiah for the pilgrimage costs.

Politically, the *iri* have achieved what they consider to be a better situation. Although all important positions are still exclusively in the hands of the *mel*, the *iri* have been able to participate in the most important political event in the village—the election of the modern village head. This has become possible because the election is now based on Indonesian government rules under which social rank is not taken into account. These rules state that every villager who is at least 17 years old (or younger if married) can participate in the election of the village head and parliament members.

It was also apparent that the *iri* had gained courage to strive for more freedom, or at least refuse unfair treatment by the *mel* through political participation. For example, many *iri* families, primarily from the Muslim settlement, actively worked towards abandoning the 'traditional' relationship they had with their masters. To this end, in Nuhuyanan some *iri* voted for the village head candidate from Rahaded instead of the candidate from Nuhuyanan *fam* because of his promise that if he won the election, he would eliminate the boundaries between social ranks. At the Christian settlement, a serious conflict between *mel* and the *iri* took place in 1987 triggered by the elopement of a *mel* woman with an *iri* man. The *mel* woman's family did not accept the relationship. They took the woman back, beat the man, and brought the case before the village head. The man was fined by the customary court but neither party was satisfied that the issue was adequately resolved. The man's party, supported by other *iri* families, considered the decision unfair because in previous cases where a *mel* men had eloped with *iri* women, the men were not beaten or fined as heavily. The *mel* believed that this case had been pursued with the specific purpose of challenging their domination at the Christian settlement. Although the case was formally closed, each party still retained ill feelings towards the other. These tensions hampered communal working relations and surfaced with the building of a new church. When the *mel* woman's family was working, a relative of the *iri* man came to collect some tools so that his team could start making bricks, but the woman's family did not allow him to take them. The man's relative explained what had happened to his team, who by chance were mostly *iri*, and they interpreted these actions as a rejection of their involvement in constructing the church. Again, conflict was ignited. A church commission from the central missionary came in to settle the situation with a novel diversion—a challenge to all families to construct a fence around, or at least in front of their houses. Interestingly, the *mel* constructed brick fences while the *iri* erected wooden fences.

Village Leadership

Everyone in Dullah Laut seems to agree that the Henan *fam* formed the first settlement on the island.[10] As the first inhabitants they were entitled to hold the position of landlord. This title indicated that they were the owners of all the land and adjacent waters (in this case, the island of Dullah Laut and its waters). As holders of the position of landlord, they had to be consulted whenever members of the community intended to make a new garden. The landlord was also considered to have a 'spiritual' attachment with the land and to be its guardian. Therefore, he was needed not only because of his position as the owner of the land but also because he was the only one who could communicate with the invisible owner from whom spiritual permission should be requested.

Unfortunately, this *fam* has disappeared.[11] There are two versions of the story accounting for this *fam*'s disappearance. The first concerns their physical appearance. The people of Henan were believed to be short with elephant-like

---

10 Riedel (1886: 218) noted that the first inhabitant of Dullah Laut Island was born from an areca nut flower. However, he did not mention that it was the Henan *fam*.

11 A similar narrative stating that the real landlord disappeared was told in Ohoitel (Laksono 1990: 101).

ears and as more people came to settle on the island, the Henan were driven inland to the forest finally isolating themselves by becoming 'invisible people' (*orang ilang-ilang*).[12] The second version asserts that those who were driven out were only the lowest rank Henan, while the *mel* Henan vanished simply because they lacked members.

This narrative is very important for other *fam* in Dullah Laut. This is because when the real landlord disappeared, Dullah Laut became an 'unclaimed' island meaning that the door was opened for other original *fam* to stake a claim. The narrative has also provided a new direction to the discourse of precedence based on the assumption that since the original Henan no longer exist, the landlord position should be left out of the discussion of who has rightful claim of ownership of Dullah Laut. The discussion of ownership of Dullah Laut has now turned into a discourse on political leadership based on a common belief that most of the landlords in Kei had transferred their rights to the leaders of the villages.[13]

The claim to being the first immigrant on Dullah Laut was proposed by two pairs of *fam*: Henan(2)-Rahaded and Yamko-Lumevar.[14] Their claims were based on the narratives of their ancestors. The Henan(2)-Rahaded *fam* narrative starts from the village of Har on the east coast of the northern part of Kei Besar Island (Map 1-2). It was said that a wild dragon that ran amock had driven the inhabitants out of the village. To avoid the danger, two of the villagers, Bad and Sam, departed their homeland and sailed to the north. After passing Tanjung Burang, the northern-most cape of the island, they turned to the southwest. Finally, they anchored on the white sand of Dullah Laut beach. According to the second version of the first narrative fragment, Bad and Sam were accepted by Henan. Bad lived there and developed his own *fam* called Rahaded.[15] Sam, due to a lack of male members in Henan, was adopted as a member of that *fam*. When the real Henan died out, Sam continued holding their *fam* name.

For the Henan(2) and Rahaded *fam*, this piece of the narrative is clearly considered to be proof of their precedence. As the narrative suggests, their ancestors' arrival on Dullah Laut was accepted by the first Henan—the real Landlord of Dullah Laut. Furthermore, the adoption of Sam by the first Henan not only strengthened this association but also demonstrated a special

---

12 These invisible people are not spirits or ghosts, according to local legend. They are real human beings but for some reason they isolate themselves in an invisible world. However, they are believed to be immortal and powerful.

13 Van Hoëvell (1890: 132) noted this transfer had occurred just a few years before he travelled to these islands in October and November 1887.

14 The Henan in this narrative is the second Henan, the first Henan having vanished (see the beginning of this section). I will refer to the second Henan as Henan(2).

15 Members of Rahaded *fam* interpret the term 'Rahaded', which derives from the words *rahan* and *ded,* as the 'pioneer house'. Others understand it as 'the house that is opposite on the street'.

relationship between the two parties. An old Henan(2) member told me that an adoption would not be conducted except for those who had a kin relationship or were considered to be very special.

The Henan(2) position is also supported by another fragment that recounts the story of Raharusun *fam*. The ancestors of this *fam* were believed to come from Luang. They migrated to Langgiar Fer on the southern part of Kei Besar Island, and then moved to Tetoat on Kei Kecil Island (Map 1-2). A diarrhoeal epidemic then forced them to leave Tetoat.[16] Sailing to the north, the Raharusun ancestors anchored in shallow waters in Dullah Laut territory. To avoid the epidemic, they lived on their wooden prau for three months.[17] Finally, they were found by the Henan(2) *fam* members, called Sertut Renan and Sertut Yaman, who invited them to join them on Dullah Laut. Assuring them of their good intentions, Henan(2) presented the island of Moa as a gift to Raharusun. Their conversation was recorded in a traditional song that goes as follows:

Henan	It yaaw waruh mehe at bahaok umat antal o dan be at her ardofa, it tes atdok did nuhu tanat, nuhutanat ohoi Duroa.	We are the only two persons, looking for other people to live together on our island and land by the name of Dullah Laut.
Raharusun	Ooo am yaaw wartil am takloen amtav nuhu Tetoat, we lo mama amba haok mang rir nuhu atau tanat am her vo amnes amdok am ames ohoi nuhu ain mehe.	We are the three brothers, moving from the land of Tetoat, looking for others' islands or lands where we can join to live together.
Henan	Am her il imdok fo ites at dok famehe yu nau amna mo mam nuhu tanat oo mam ohoi meman Duroa ooooo.	We ask you to come to live together on our island/land and the hamlet by the name of Dullah Laut.
Raharusun	Im bir ohoi bir woma naa te waaed ooo?	Does your hamlet have a centre?
Henan	Woma meman Varne Harmas oooo.	The centre is named Varne Harmas ooo.
Raharusun	Bir ngur meman aka oooo?	What is the name of your sand?
Henan	Ngur Lak Laver ooo	Ngur Lak Laver
Raharusun	Bir tahait meman aka ooo?	What is the name of your sea waters?
Henan	Tahait Sir Dabro	Tahait Sir Dabro
Raharusun	Bir nam meman aka ooo?	What is the name of your deep sea?
Henan	Nam Ngil Ublay ooo	Nam Ngil Ublay

---

16   Van Hoëvell (1890: 153) found that epidemics frequently forced people to move out of their village and establish a new village elsewhere.

17   In remembrance of this event, the coastal waters were called Ibun vuantil (*ibun* meaning sea beds/grass, *vuantil* meaning three months).

This traditional song is considered to be the strongest 'proof' of the narrative. What does this song prove? Henan was an influential person, as shown by his invitation to Raharusun. Only those who hold privilege may invite others to live with them, a notion which is bolstered by the story that Raharusun was offered Moa Island. Henan's answers to the questions regarding the customary names of his territory are additional proof of his privilege because only prominent persons can master the customary names of their territory. Finally, the song also implies centrality as it associates Henan with Woma Varne Harmas, the centre of the village. In this context, the Henan then claimed that their ancestor was the chief of the village.

The oral history of Yamko[18] *fam* proposes another claim on Dullah Laut. Conflict between the *mel* and the *iri* had forced Varne and his wife—the ancestors of this *fam*—to move away from Uf on Kei Kecil Island. They found that Awear on the northern part of the island was a convenient place to live, so they settled there. They had two children—the first was born with gold teeth and became a goldsmith. The place was given a name after his profession, Vaan Fomas (*vaan* meaning cave and *fomas* meaning goldsmith).

The story goes on to introduce the ancestors of another *fam* and its relation to Yamko's. Driven from the island of Banda, Kabir—the ancestor of the Nuhuyanan *fam*[19]—landed at Wada Iyuwahan[19] on the northern part of Dullah Laut, near Awear (Map 5-2 12). Because the land was not habitable, he continued his journey to the island of Wara Fangohoi, which later became known as the island of Rumadan (*Rumah Orang Banda* meaning the house of the Bandanese). Varne found Kabir and invited them to join them at Awear. Kabir agreed and Varne arranged for him to marry his daughter, Lumas.

Starting from this point a settlement took shape. Led by Varne, his and Kabir's descendants established a hamlet. First, it was located around Vaan Fomas. Later they moved inland to a place called Tenantua Ohoi. As the population grew, Tenantua Ohoi became too small. Finally, they moved some kilometres to the south and erected another hamlet. Yamko's ancestor was recorded in this place and his name, Varne, was used as the name of the settlement centre.

This fragment is used to legitimize the privilege of the Yamko *fam* over Dullah Laut Island as well as Yamko's superior position over that of Nuhuyanan. The name of the village centre, Woma Varne, is strong proof of their privilege. They argued that the settlement centre was intentionally given the name of their

---

18   There was an ancestor of another *fam* who came to Dullah Laut together with the ancestor of Yamko's. This was the ancestor of Lumevar *fam*. They were considered to be a brother *fam* (*fam adik-kakak*). This is why in ritual idiom Yamko is always associated with Lumevar. However, this *fam* vanished and no one could remember any part of the story narrating their history.
19   A shell-like stone at Wada Iyuwahan is considered proof of this story. People believe that this stone was Kabir's vehicle which transported him from Banda to Dullah Laut Island.

ancestor because of his position as a great man and head of the hamlet (*hilaay*). The creation of Vaan Fomas and Tingivan are further proof of their precedence. Tingivan is a face-like relief on a stone at the beach with the same name. The relief is believed to be the face of one of Varne's servants.

There is another fragment that mediates these two claims. This fragment acknowledges Henan-Rahaded and Yamko-Lumevar as the first two immigrants to come to Dullah Laut since both settled and controlled different parts of the island. Henan-Rahaded and their allies Raharusun-Rahawarin, who were called Ohoiroa, occupied the southern part of the island with Woma Hermas as its centre. Yamko-Lumevar and Nuhuyanan-Songyanan, the Fauur people, controlled the northern part of Dullah Laut, with Woma Varne as its centre. Before they met, each developed their own people and maintained their own territory. Once they realised the existence of the other, boundaries were erected between the two groups. According to some local elders, this boundary was a stone fence stretching from the northern edge of the present Christian settlement in a westerly direction to Foarne Ohoi, to a spot near the island of Moa.

Marriage contracts between *Ohoiroa* and *Fauur* members and the frequent occurrence of war in the region encouraged these two groups to merge. For this reason they abandoned their previous settlements and constructed a new one at the present Christian settlement. This was the crucial point in Dullah Laut history because, starting from this point, rights and obligations were shared among the eight *fam*. This meant that all decisions concerning Dullah Laut as a community had to be decided by meetings comprised of representatives of all origin *fam*, that is the *Ohoiroa Fauur*. In return, all members of *Ohoiroa Fauur* were responsible for defending their territory from outside intervention. This is seen in the membership of 'the thirty troops' of *Ohoiroa Fauur*, the group of 30 traditional soldiers representing the original *fam* of Dullah Laut who are responsible for defending the territory.

This narrative has shifted the discussion of leadership, which was formerly based on the issue of the first settler—a 'founder focused ideology', borrowing Bellwood's term (1996)—to a 'personal achievement' ideology. By 'personal achievement' I mean the role of members of a particular *fam* in events that were crucial for the Dullah Laut community. This alternative discourse has raised the position of non-first-settler *fam* from a subordinate to a more equal position. Some cases even demonstrate the precedence of non-first settler *fam*. I will relate a narrative that demonstrates this point.

Bal Ulab Nuhuyanan was a guardian of the law which says: '[those who] paddle should paddle with the sharp side of the paddle, [those who] use a stick [to move their prau] should use the stick upside down, [those who] bail out [the prau] should use the back side of the container' (*an vehe an hov vehe ngoan, an leak an hov leak tutu, an it vaha an hov yer tetan*). He was very strict in imposing this law. No culprit escaped his sword and his agility with the weapon was the reason he was called 'the lightning from the north' (*anvitik sarab ribat naa bad maar*). Arnuhu, the King of Danar, considered him a dangerous enemy but at the same time needed him as an ally. The king sent a moon-shaped medal (*mas a yam vot*) to Bal Ulab Nuhuyanan, nominating him to be a king of his region. Because Duroa [Dullah Laut] was a very small island, Bal Ulab Nuhuyanan rejected this nomination. Eventually, King Danar appointed Baldu Wahadat, the leader of Dullah Darat village on Dullah Island, as the new king.[20]

This narrative clearly shows how the achievements of Bal Ulab lifted the social standing of Nuhuyanan within Dullah Laut and led to him being considered a local leader.

Territory

While these narratives discuss particular issues of territory they do not apply to the whole of the Dullah Laut territories (*petuanan*). The narratives deal only with Dullah Laut Island, which is one of nine islands that comprise the village territory. The following narratives provide reasons for the incorporation of other islands into Dullah Laut territory.

The narrative begins with the story of Utan Fak Roa (*utan* meaning a group of settlements or village, *fak* meaning four, *roa* meaning sea). As the name implies, they were four hamlets located on Dullah Laut, Ohoimas, Rumadan (Warohoi Island) and Ngang Hangar Laay (Map 1-2).[21] Those who lived on Ohoimas Island were known as the Ohoimas people.[22] Their territory covered not only the island where their hamlet was located but also the islands of Baer, Sua, Watlora, and their adjacent waters. On the island of Rumadan lived the people of Wara, Fangohoi, and later immigrants from Banda Island. Their territory covered both the islands of Rumadan Warwahan and Warohoi and their adjoining waters.[23]

A conflict called the 'Waterspout War' (Vuun Asnen) broke out on Warohoi Island. The disputing parties were the Wara and Fangohoi against the Rumadan people who originally came from Banda. They fought over a waterspout that was used as a rainwater collector. The people of Rumadan, supported by Bal

---

20 The former King of Ibra, Moh, Fagi Renwarin (Renwarin: n.d.) and Ohoitimur (1983: 55) wrote similar narratives. The present King of Dullah rejects these versions.
21 People believe that Ngang Hangar Laay has been submerged and is now under the sea.
22 Riedel (1986: 215) noted that Suwa, Ohoimas, Baer, and Watlora islands were attached to Letman Village.
23 Riedel (1886: 215) found these islands were a part of Tamadan village territory.

Ulab Nuhuyanan,[24] drove the Wara and Fangohoi off the island.[25] The Rumadan people lived on the island until they all moved to the village of Tamadan on Dullah Island.

Because of Bal Ulab Nuhuyanan's involvement in the Waterspout War, the people of Dullah Laut claimed possession of these islands. However, the people of Rumadan contested this claim. In 1967, Tamadan, the traditional village leader, requested the people of Dullah Laut stop making use of the Rumadan Islands. The people of Dullah Laut were very upset and put signs of possession around the islands. They then reported their action and their reasons to the King of Dullah. Two days later, the two parties were called before a customary court. The court failed to determine who owned the islands, but it was decided that both parties would share equal rights to the Rumadan Islands to avoid escalating the conflict.

Claims over Ohoimas and Baer islands are based on another narrative. Once, some Papuans (Nisyaf) came to Ohoimas for the purpose of collecting human heads and raided Ohoimas Island. While some were killed, a woman called Ngirut escaped. She swam to Rumadan Island and had a rest on a beach that was later named after her, Ded Ohoimas. She continued her escape by swimming across the strait between Dullah Laut and Rumadan. She reached Dullah Laut at a place called Wear Ohoimas. After walking to the village of Dullah Laut, she reported what had happened on Ohoimas. *Ohoiroa Fauur*, headed by Yahaw Rahaded, the traditional village leader, declared war on the Papuans who were still on Ohoimas Island. The war between the two parties—called the Papuan War— broke out on the island of Ohoimas. *Ohoiroa Fauur* defeated the Nisyaf. The Papuan commander was killed in a duel with the commander of *Ohoiroa Fauur*.[26]

In 1920, the people of Letman contested this claim. They went to Ohoimas and put some signs of possession around the island. Paddling back to Letman, they passed Dullah Laut Village and sang the following song: *"aklul sang sang vat larito vat yaf o be"* ("[we are] sharp like metal, hard like stone and hot like fire, ready to face all the challenges or the enemy who comes"). This was considered to be a war challenge by the Dullah Laut people who chased the people of Letman and reported the event to the King of Dullah. This case was brought to a customary court led by the king, but it failed to reach a satisfactory decision. The people of Dullah Laut and Letman appealed and took the case to the Dutch

---

24 An old Rumadan informant who lives at Tamadan Village on Dullah Island told me that Bal Ulab was not involved in this event.
25 The descendants of these people can be found at Faan and Sathean village on Kei Kecil Island.
26 Mr A. Rahaded claimed that it was Yahaw Rahaded who fought with the Papuan commander. Mr M. Rahawarin believed that it was their ancestor by the name of Masen Yaman who was the *Ohoiroa Fauur* commander that killed the Nisyaf commander in a duel.

government in Ambon and a copy of the decision was given to the King of Dullah. The people of Dullah Laut have interpreted the court's decision as being favourable to them and conclude that Ohoimas Island should be ruled as part of their territory. Interestingly, the people of Letman support this conclusion. Although the official ruling on the case was never produced,[27] Raja Dullah claimed that the decision clearly indicated that the islands were under his control.

Conclusion

This chapter discusses the most important aspect of Kei tradition. The *toom* is both a narrative of origin and history according to the people of the Kei Islands. *Toom* is the only point of reference when people talk about tradition, therefore understanding the *toom* is the only way to understand tradition.

The *toom* that explains the construction of society form the foundations of Kei tradition. According to the *toom*, the Kei people are divided into three main social ranks: the *mel*; the *ren* or free people; and the *iri*. Based on this *toom*, relations between social ranks are defined based on political and territorial domains. Other *toom* describe the formation of a specific domain such as a kingdom, village, or settlement. Again, these *toom* provide the basis for explaining how a particular group of people came to be connected with a particular territorial and/or social domain.

At the practical level however, the explanation of relations between the *mel*, *ren,* and the *iri*, as well as relations between social groups within a social rank is not as simple as it is described by a *toom*. This is because the same *toom* is open to different and often contradictory interpretations. Frequently there is more than one *toom* that explains the relationships between social groups in a single domain. This means that different social groups may propose different forms of relations and put forward different claims by drawing on the same or different *toom*.

The multiple interpretations of *toom*, as well as the existence of multiple *toom* describing a particular issue, could be interpreted as reflecting the flexibility and richness of tradition. However, as with newly introduced structures such as religion and politics (see Chapter Two), when people consider their interests to be more important than those of others, the *toom* serves merely as a vehicle to promote those interests. Using a motor sport analogy, the *toom* could be seen as a display of racing cars. People choose the most efficient and effective car and

---

27   I intentionally added this information to show that even for an event recorded historically, the written proof is not considered as important as the oral history.

if they cannot find one, they make a better car. In this circumstance, winning the race is more important than the car itself. In this sense, the *toom* became an object of history rather than representing history itself. It is the people—driven by their interests—that create the *toom*, rather than *toom* determining how people should behave.

This logic is very apparent if we look at how the *toom* was used in conflicts relating to the political and territorial control over particular domains in the Kei Islands. An analysis of the conflicts that occurred in the Kei Islands forms the basis of the following chapters.

5. Land and Sea Tenure in the Kei Islands

As mentioned in the introductory chapter, the focus of this book is the politics of communal marine tenure in the Kei Islands. While the preceding chapters set out a broad context of the Kei people, this chapter will provide a specific understanding of communal marine tenure. The knowledge of marine tenure's general principles and practice will be crucial in understanding the politics of marine tenure that will be elucidated in the next four chapters.

Two notes worth mentioning before I start explaining customary marine tenure. First, I will make comparisons between general principles and the practice of marine tenure in Watlaar Village on Kei Besar Island (see Rahail 1995) and Dullah Laut Village. This comparison is intended to show that even within villages in the Kei Islands, the general principles of traditional marine tenure are not always the same. Thus even in a single cultural group, the understanding and practice of marine tenure varies. Second, this chapter will also illustrate the flexibility and contestability of the principles and practices of marine tenure which may be different from our understanding as they relate to modern concepts of rules and regulations. These practices will explain the nature of conflicts that I will discuss in the following chapters.

The Concept of Territory

The basic concept of territoriality in the Kei Islands is embodied in the concept of *petuanan*. The word itself derives from *tuan* which means 'owner' or 'master'. The prefix *pe* and suffix *an* attach a specific location to this word. Therefore, it is not far from its literal meaning for the people of the Kei Islands—and for many other communities in Maluku (Zerner 1992)—who understand it as an estate or territory of a certain traditional social group.

As islander communities, the concept of territoriality covers ownership of both land and sea. This linkage is expressed with the pairing of *petuanan* with terms that refer to sea territories such as: sea estate (*laut*); coastal area (*met*); and sea (*roa*) and land territories such as: land estate (*darat*); island (*nuhu*); and land (*nangan*).

For specific purposes the concept of sea territory can be discussed independently from land territory, although it would be difficult to comprehend the former without any reference to the latter. This is because, to some extent, the concept of sea territory is an extension of land territory. In other words, some elements of the concept of the sea territory derive from the land. For this reason, I will discuss land territory before dealing with sea territory.

Land Territory

The concept of land territory in Dullah Laut refers to the nine islands of their territory. These islands are Dullah Laut (Duroa), Moa, Adranan (Dranan), Rumadan Warwahan, Rumadan Warohoi, Sua, Baer, Ohoimas and Watlora (Ruin) (see Map 5-1). When talking to outsiders, Dullah Laut villagers consider these islands to be inseparable. Thus when a villager says 'this is my island and land' (*nii yaaw nuhu tana*), he or she is talking about all the islands of the Dullah Laut territory. And of course, when a person is talking to an outsider, the word *yaaw* (meaning 'I') does not refer to that person as an individual but as a member of the Dullah Laut community. A different way of addressing the issue of land territory would be used with fellow villagers. In this context villagers will divide each island into zones based on use.

The following identification of land territory was set out by Rahail (1995: 18–21). Rahail is the ruler of Maur Ohoi Wut Kingdom on the northern part of Kei Besar Island (see Map 1-2). He is a well known traditional leader, particularly among NGOs. He has written two books concerning tradition in his domain (Rahail 1993, 1995).

Rahail (1995) divides the land area into several zones (Figure 5-1). The first zone, closest to the sea, is where the settlement (*ohoi*) is located. Apart from housing, people use this zone to grow decorative and medicinal plants and tend domestic animals such as chickens, goats and pigs. The second zone is primarily a zone of intensive cultivation (*ohoi murin*). Vegetables, spices, cassava, corn and nuts are planted in this area. Because this area is still relatively close to the settlement, people use it for their domestic animals as well. According to Rahail, the third zone (*rok*) is similar to the cultivated zone but with plantings of other crops and fruit trees. The fourth zone is the area where people practice shifting cultivation (*kait*), and the highest part of their territory is the zone covered with forest (*warain*). People cut trees for house construction and hunt wild animals in this zone.

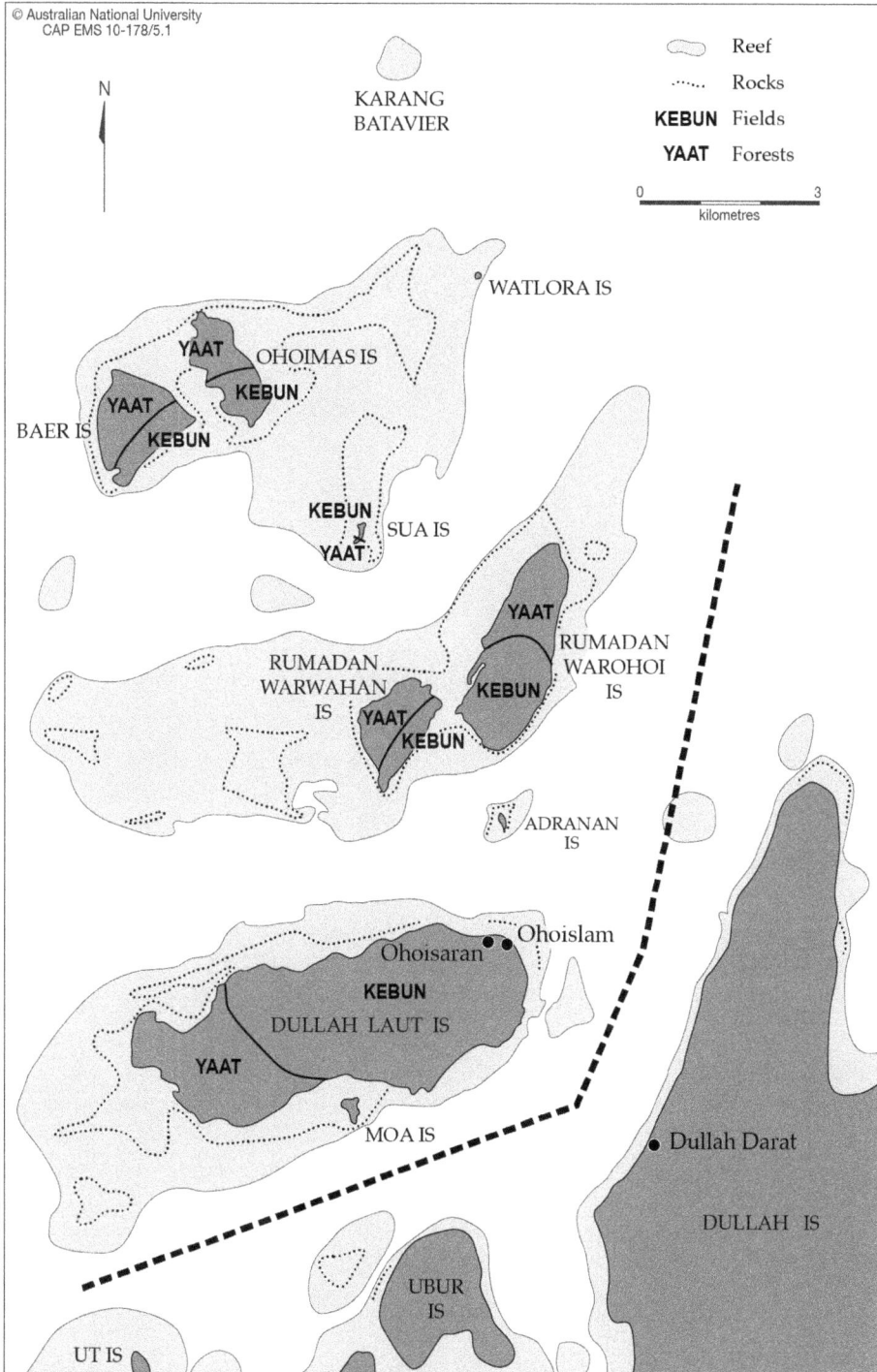

Map 5-1: Dullah Laut island territories and land classification.

Source: Author's fieldwork.

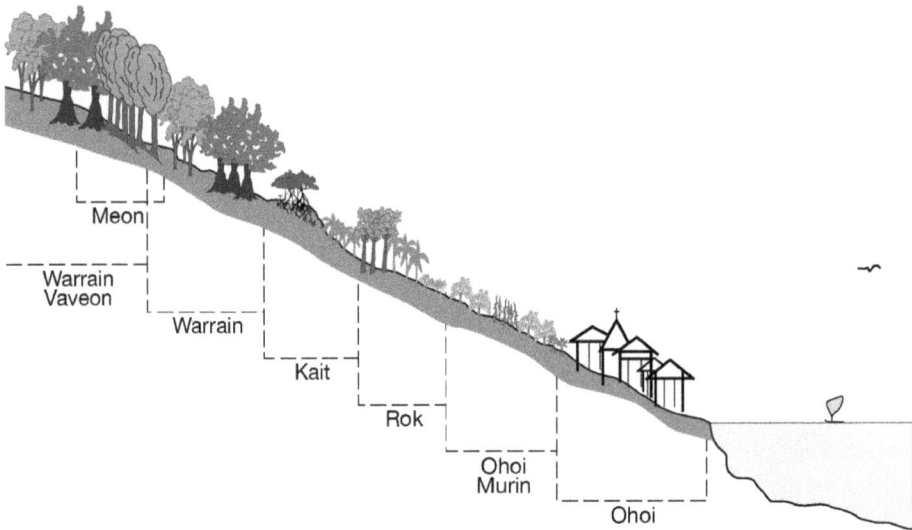

Figure 5-1: Land zone classification of Watlaar, Kei Besar.

Source: Adapted from Rahail (1995: 21).

The division of land territory is more simply defined in the Dullah Laut territory (refer to Map 5-1). The discussion of land territory among the villagers is more clearly represented by the word 'island' (*nuhu*). In Dullah Laut, the island is divided into four sections. The first section refers to a plot of land where a house and its garden are located (*kintal*). The aggregation of house plots form a hamlet or settlement (*ohoi*)[1] and further inland is the area where a range of cultivation activities take place (*kebun*). The concept of *kebun* in Dullah Laut encompasses three distinct zones recorded in the Kei Besar land zone classification: *ohoi murin; rok;* and *kait* (as shown in Figure 5-1). This indicates that people in Dullah Laut do not differentiate between areas which are used intensively and those for shifting cultivation. However, people do use the expression 'unused garden' (*bekas kebun*) to distinguish areas where cultivation activities have stopped either temporally or permanently.[2] At the highest altitude of the island is the forest (*yaat*). This area is the only zone in which people's activities are limited.

Looking at these divisions on each of the nine islands, it is possible to distinguish that Dullah Laut is the only island where all four zones of land territory exist (refer to Map 5-1). On this island dwell the Ohoislam (Muslim) and Ohoisaran

---

1 It is important to note that the meaning of *ohoi* here refers to a space or territory, which is different from the political concept of *ohoi* as a social organisation.

2 The term 'permanently' here refers to a garden that has been abandoned for more than the usual fallow period (one to five years).

(Christian) settlements (*ohoi*) with their gardens (*kintal*) in close proximity. The cultivated region (*kebun*) covers the widest area of the island. Except for a small portion of forest (*yaat*) on the western tip of the island, the two settlements, and some sacred areas, the whole island has been cleared for cultivation.

As for the distribution of land on the other islands, Sua Island has mostly been used for cultivation and a very small portion of the island has been left undisturbed to support the forest. Rumadan Warohoi Island is divided into cultivated and forest areas in almost equal portions. Baer, Ohoimas, and Rumadan Warwahan islands have a small portion of cultivated land with a greater area of forest. Watlora Island is entirely forest because people consider it to be too far from their settlements for cultivation purposes. Adranan Island is very sandy in nature and even mangroves grow sparsely. People have tried to plant coconut trees but they did not grow and the island has been left uncultivated ever since. The mangroves that do grow on the island are not considered a forest and people call this island 'an empty island' (*pulau kosong*).

Sea Territory

To understand the concept of the sea territory, I will refer again to Rahail's classification of Watlaar on Kei Besar Island. He divides the sea territory into ten zones (see Figure 5-2). The first zone includes the coastal area which is dry during the lowest tide (*ruat met soin*) and has a maximum depth of around three metres. The second zone stretches from the minimum to maximum low tide (*met*) with a depth of three to five metres. The third zone is the frontier between the tidal zone and the deep sea (*hangar soin*). This zone is never dry, even during the maximum yearly low tide season in the last months of the year. According to Rahail (1995: 23), this area is covered by coral reefs that sustain a wide variety of fish species, molluscs and other sea organisms. The depth of water in this area ranges from five to fifteen metres and this is where villagers set their small fish pots and gill nets and use lines and spears to fish. The next zone is deeper and may reach up to 100 metres (*nuhan soin*). The ecological characteristics of this zone are similar to the previous zone but the coral is larger and much more diverse. The next zone is identified by the dark blue colour of the sea (*faruan*), reaching 100–200 metres in depth and traditionally used for tow fishing[3] and setting fish aggregating devices. The outer five zones are as follows: *wewuil; wahdaan; leat dong; walaar entetat;* and *tahit ni wear*. These names are derived from a reference point on the land visible from the zones. In discussing these divisions, Rahail claims that the zone that is furtherest from the

---

3   Involves a fishing line towed from the back of a canoe.

land is the frontier of the people's sea territory. This means that the area from that furtherest zone to the innermost named zone is under their control while the unnamed zone beyond is not available to them for fishing purposes.

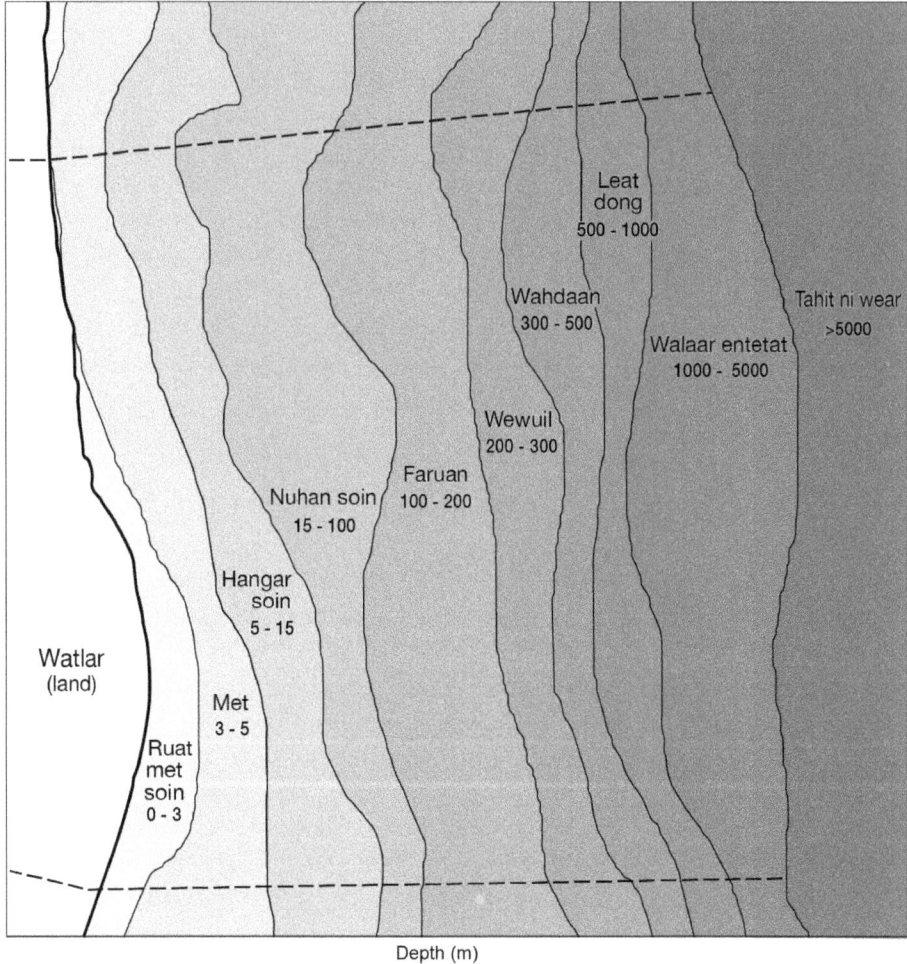

Depth (m)

Figure 5-2: Sea territories classification of Watlaar, Kei Besar.

Source: Adapted from Rahail (1995: 22).

The concept of sea territory in Dullah Laut is simple compared to Rahail's model. Unlike land territory, villagers consider their sea territory to be inseparable. There is only one word representing their concept of the sea territory: *meti*. This word expresses their main claims regarding sea territory.

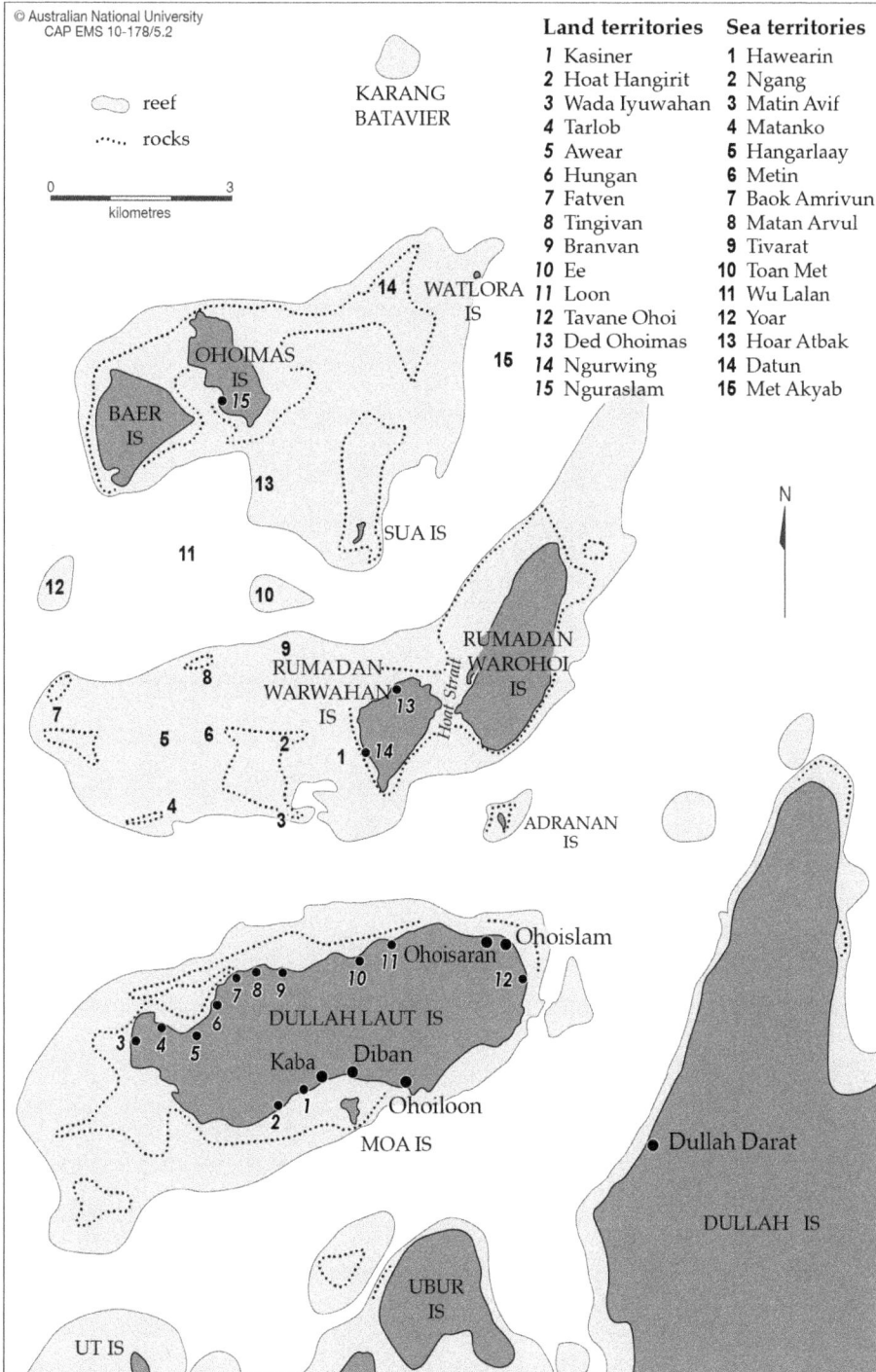

Map 5-2: Land and sea territories of Dullah Laut.

Source: Author's fieldwork.

Note: *Kebun* refers to fields and *yaat* refers to forests.

The term *meti* in the context of sea territory is broadly understood in two different ways.[4] The first refers to the coastal area from the high tide mark to the boundary of shallow and deep sea water. It also refers to submerged atolls and underwater reefs. The first version of *meti* is used when people describe the sea territory surrounding and within the nine islands of their land territory. The second meaning describes submerged atolls and underwater reefs in the area outside of the first. The location of this *meti* might be miles away from the nine islands such as the 'snake meti' (*meti rubay*), located some three to four nautical miles to the north of Baer Island.

The detailed naming of *meti* compensates for the lack of spatial division of the sea territory. The people have developed quite detailed names for their sea territory. As can be seen in Map 5-2, people even divide a block that might be considered a single *meti* into smaller parts and give each a different name. I believe this is done because a name is very important since it is proof of a connection between people and a particular place. As mentioned in Chapter Four, names are also proof of the 'historical truth' of a particular narrative by which people may claim special attachment to a place.

Rights over Territory and their Distribution

Two rights attached to a territory define the exclusivity of tenure. The first is the 'use right' (*hak makan*). The holder of this right is entitled to make use of the territory and can exercise the use right by cultivating a plot of land, cutting down the trees in the forest, and fishing. The second is 'property right' (*hak milik*) and it confers greater powers than use right. Those who hold this right are not only free to make use of the territory, but can also freely transfer their use right to another party. A contract between the committee of origin group (*Ohoiroa Fauur*) under the leadership of the village head granting permission for a fishing company to fish in the Dullah Laut sea territory is an example of how the property right holders transfer their use right to another party.

The distribution of these rights differs depending on the context of the discussion. As with the division of land territory, who is involved in the discussion influences how people address the issue. Also, the distribution of rights differs depending on the specific property concerned, such as whether it is village territory, or land or sea territory. The following discussion will deal with this issue in more detail.

---

4   In addition to these two meanings, *meti* is also understood to mean low tide. For example, when someone asks '*So meti ka seng*?' the person is asking 'Is it low tide already?'

Traditional Domain as the Right Holding Unit

When I started this discussion on issues of territory, I defined *petuanan* as an estate or territory of a certain social group in a traditional domain. Now, I would like to detail the units of traditional domain that control a particular territory. This is an important issue because when people talk to outsiders, it is these units to which issues of ownership refer.

In the Kei Islands, various traditional domains control territorial tenure. An example of this is the settlement *petuanan* (*petuanan kampung*) of Hollay and Hoko on Kei Besar Island (Map 1-2). Each of these settlements controls its own territory, despite the fact that they are administratively a part of Hollat Village. Some territories are controlled by a larger domain, namely the village (*ohoi* or *desa*). Dullah Laut is an example of this. Although Dullah Laut Village consists of two different settlements, they control a single undivided territory. Thus, whenever they have to deal with issues of territorial tenure with outsiders, the village will handle it. Another traditional domain that in some cases controls a single territory is a kingdom, for example the Kingdom of Ibra on Kei Kecil Island (Map 1-2). This kingdom— comprising the three villages of Ibra, Sathean and Ngabub—controls a single territory. This means that none of these villages has the autonomy to handle issues relating to their territory and all three are under the leadership of the King of Ibra who can speak for their territory. I was also told that there were some territories which were controlled by the nine group or the five group (see Chapter Two) although I was unable to identify the territory referred to. Finally, there is a territory that is called 'public territory' (*petuanan umum*) that is shared between the nine and five group. There are two views concerning the exact location of this public territory. Some suggested that it was Daar Island and its surrounding sea in the Kei Kecil Archipelago. Others believed it was a submerged island, located some kilometres to the east of Daar Island.

Before I discuss the distribution of rights over territory within the domain, I will return to Dullah Laut to look at how the village represents the interests of territory right holders to outsiders. I suggested earlier that in Dullah Laut the village controls the territory whenever they have to deal with outsiders. There are two types of scenarios where the village exercises this role. The following two examples illustrate these scenarios:

1. In July 1996, a fisherman came to the village leader's house. After being accepted by the village head, this fisherman introduced himself. He was from Madura (see Map 1-1), the owner and skipper of a boat which was fishing for sea cucumber in the Kei Islands. He asked the village leader if he would be allowed to search for sea cucumber in the village sea territory for about a week. In return, he would give an amount of Rp50 000 (approximately US$22)

as contract money. The village head agreed and wrote a letter of permission. Handing over the letter, he then told the fisherman to show the letter if any Dullah Laut villagers questioned his fishing activities. I saw the boat passing the village the day after. I also noticed the boat anchored near Ohoimas Island (Map 5-1) some days later.

2. In October 1996, Mr JW came to the village head's house. Mr JW was a representative of a grouper fishing company operating around the Kei Islands. His company was involved in reef fishing and trade and their business was not a small one, according to local perceptions. After some introductory remarks, Mr JW said that he intended to make a contract with the village concerning his intention to bring his ship and fishermen to fish in Dullah Laut territory. The contract he talked about was for a period of a year.

The village head agreed in principle. However, he said that the issue of contracting sea territory to a fishing company was beyond his authority and explained that he would call a customary court meeting attended by representatives of the origin *fam* from both settlements in the village as well as Mr JW or another person representing the company. This customary court would discuss whether the company would be permitted and if so, how much money they would be asked to pay for the use of the territory. The village head then made an appointment with Mr JW to hold court a week later.

The customary court was held as planned. Each representative of the origin *fam*, *Ohoiroa Faur*, was given a chance to express an opinion on the issue. Finally, the court decided to sign the contract. The company was allowed to operate in their territory and in return they had to pay Rp3.5m (approximately US$1522). The money was distributed to *Ohoiroa Faur* representatives, each receiving Rp30 000 (approximately US$13). The rest was allocated for the construction of the village mosque and church.

Some points should be made to clarify these two examples. Both examples concern the transfer of rights over the village sea territory, which means that both also concern property rights issues. The first example is a scenario in which the village head alone represents the interests of the right-holding unit. The second scenario illustrates a situation in which transferring the right over sea territory is beyond the village head's authority. In this case, the property right was put in the hands of the 'committee of origin *fam*' (*Ohoiroa Fauur*). I should also note that the application of either scenario depends on the villagers' consideration of the level of exploitation by the outsider. The level of exploitation by the Maduran fishermen was relatively low over a short period, while the grouper fishing company was considered a big enterprise in which a high level of exploitation for a lengthy period was to be predicted.

Distribution of Rights over Land Territory

The rights over land territory are distributed to different units within the village. The section of the territory influences how the rights are distributed. Rights to each house plot are granted to the family or families who constructed a house on it. Families might obtain this right from their parents. Theoretically, the property right is in the hands of the parents. When they die it goes to their son(s), while their daughter(s) get only use right. Other than inheritance, rights to house plots can be transferred from one family to another through trade, exchange, or gift.

The principal premise of the distribution of property rights to a field of mixed annual crops (*kebun*) is that the right is given to those who clear or open the forest and cultivate it. Based on this premise, the property right holder of a field ranges from an individual to a particular segment of a *fam*. In the case of an individual property right holder, this could refer to a new field that has been made by clearing the forest for example, the fields that were newly opened at Awear on Dullah Laut Island (see Map 5-2). With *fam* member property right holders, this refers to fields that were made decades ago. The property right on these fields has been transferred for generations. Theoretically, the more ancient a particular field, the larger the right-holding unit might be. For example, if Bal Ulab had cleared a portion of forest and cultivated it, the property right of his field is theoretically inherited by his male descendants. The number of use right holders of his field might be even larger since this right is transferred not only through the female line but also to those who are connected to the property right holder through affinal relationships. In practice however, this field might have been divided into smaller plots and the right-holding unit might be a limited number of families. This might happen because it is possible for a particular family to ask for, buy or exchange part of the field meaning that the right-holding unit of a particular part of the field is in the hands of that particular family.

The distribution of rights to forests represents the original ownership pattern of the territory. The issue of property right holding units to forests goes back to the narrative of origin. Referring to particular segments of this narrative, there are three social units that claim ownership of forests. The first social unit comprises the origin *fam*, *Ohoiroa Fauur*. Their claim is based on the narrative of origin which explains that the entire domain of Dullah Laut was aquired and developed by these origin *fam*. This segment of the narrative is interpreted as meaning that the land territory of Dullah Laut is inseparable and that each of the origin *fam* contributed equally to the occupation and development of the domain. Thus the whole territory is owned collectively by all origin *fam*. In this context, since property rights to other sections of land territory have been distributed to the smaller units, this means that references to land territory are only referring to forests. The second social unit is Rahaded *fam*. Their claim is based on the segment of narrative concerning the leadership of the village during the Papuan War and the court decision during the Dutch period (see Chapter Four). According to Mr A. Rahaded, it was his grandfather's father that led the people of Dullah Laut to win the Papuan War. Thus, he claimed, the islands of Baer and Ohoimas (acquired due to their success in driving away the Papuans), should be under Rahaded *fam* control. The Dutch also granted these islands to Mr Z. Rahaded when people from Letman Village contested their claim in the 1920s.

The third social unit that lays claim of ownership over part of Dullah Laut territory is Rumadan *fam* in Rumadan Village on Dullah Island. This *fam* claims ownership of Rumadan Warwahan and Warohoi islands. Their claim is based on the narrative of the War of the Waterspout (see Chapter Four). They believe their ancestor drove away the enemy and successfully controlled these islands and as a result, the right of ownership over the islands is in their hands.

The distribution of use right in relation to forests has never been a problem. This is because use right is not only acquired by inheritance through the male line but also through affinal relationships with the property right holding unit. In these two ways, it is assumed that all villagers— regardless of their *fam* membership—are entitled to use right. Even outsiders might claim use right if they are able to show their affinal relationship with the property right holder.

It is worth mentioning that the origin *fam* (*Ohoiroa Fauur*) and all of the villagers of Dullah Laut acknowledge the use rights given to the people from Ut Island (Map 5-1). Theoretically, this use right is limited to taking a kind of mangrove. This acknowledgment is based on the narrative of the Papuan War. According to this narrative, the people from Ut supplied weapons to the people of Dullah Island, and one of the weapons was used to kill the leader of the Papuans. To show their gratitude for this help, the people of Dullah Laut granted the people from Ut Island use rights on a kind of mangrove that grows on Dullah Laut territory. This mangrove is used as fire wood for their blacksmith activities.

Distribution of Rights over Sea Territory

The distribution of rights concerning sea territory is the same as for forests. However, besides the origin *fam* of *Ohoiroa Fauur*, Rahaded, and Rumadan, there is an additional claim to the village sea territory by the people of Ut Island. By way of background, I will discuss the three social units involved in the claim. Based on the same segments of the narrative used to claim ownership of forests, these three social units not only claim the right of ownership of the sea territory, but deny other claims as well. The *Ohoiroa Fauur* believes that the village sea territory is indivisible and thus Rahaded *fam*'s claim is nonsense. Although

Ohoiroa Fauur members agree that during the Papuan War and their legal battle with people from Letman Village that Dullah Laut was under the leadership of Y. Rahaded and Z. Rahaded, they think it was impossible for these two traditional village leaders to deal with problems without support from other members of *Ohoiroa Fauur*. In the Papuan War, there was living proof—a man called Beruntung, who was captured and then became a slave to the war commander from Rahawarin—that other members of *Ohoiroa Fauur* were involved in the fight. This meant that Y. Rahaded did not go to the war by himself. This was also the case when Z. Rahaded dealt with the Dutch government when the Letman villagers disputed their claim. Although it was Z. Rahaded who went to Ambon to meet the Dutch official to sort out the problem, the finance needed for his trip was partly provided by Nuhuyanan and other *fam* members of *Ohoiroa Fauur*.

On the other hand, the Rahaded *fam* believes that the Dutch declared that Z. Rahaded won the case and therefore granted him the right over Baer and Ohoimas islands and their sea territory. For the Rahaded *fam*, it was not the villagers as a whole who won. Based on this argument, Rahaded claims that the territories of these islands were their property. Referring to the sea territory of these islands, they insist that it is their right to give permission to—and consequently receive payment from—those who wish to make use of the territory commercially such as the Madurese fishermen mentioned in the first example in the previous section.

The *Ohoiroa Fauur* also denies the Rumadan claims on the Rumadan Islands and their surrounding waters. They believe that Bal Ulab Nuhuyanan, one of the leading figures of Dullah Laut in the past, took an active role in driving away Wara and Fangohoi in the War of the Waterspout from both islands. Therefore, since Bal Ulab was from Dullah Laut, the land and sea territories of these islands should be under the control of Dullah Laut Village.

The Ut villagers claim ownership of coastal waters called Metan Er[5] located on the northeastern side of Dullah Laut Island (see Chapter Eight). Their claim is based on their version of the narrative of the Papuan War. They believe that after helping *Ohoiroa Fauur* in this war, their ancestor was granted the property right over the sea territory. Dullah Laut villagers decline this claim because the narrative of the Papuan War referred to by the Ut villagers was not the 'right history' (*toom*) of the war. As I noted earlier, the Dullah Laut version of the history of the Papuan War only mentions that the Ut villagers were given use right over the mangroves for the firewood needed in blacksmith activities.

To conclude this section, I would like to draw attention to the fact that the exclusive right over sea territory does not apply to all situations. Theoretically,

---

5   Dullah Laut villagers claimed that the name of this *meti* is Wadaiyuwahan, the name taken from the adjacent land territory.

the sea is free to all for subsistence use.[6] People from different settlements, villages, kingdoms, or even ethnic groups are free to fish in the sea territories as long as their activities are for their own consumption. Whether they conform can be verified by the technology they use. Fishing lines, spears and arrows fall into the subsistence fishing category. In practice however, the enforcement of this norm is influenced by the personal relationship between the outsiders and local people. For example, when I went fishing in Dullah Laut territory, I often met non-villagers who were obviously, judging by their equipment, engaged in commercial fishing. When I asked the village head or other villagers if these people had asked permission to do so, I was told 'no' but that these people were either friends of theirs or other villagers, or were considered to be good people. In contrast, an informant told me that he had once driven away a fisherman from another village even though the non-villager was only fishing for subsistence. He explained:

> This person did not know his place, he did not want to share his bait with me, even when I told him that I needed only a little and did not expect it for free, I would buy it. Since I could see that he had enough bait to share, I was upset. I asked him to leave. He has not fished in Dullah Laut territory since.

The above cases demonstrate the flexibility of the exclusivity of the village sea territory. It seems that the use of the sea territory depends more on the type of relationship between villagers and non-villagers rather than the degree of the exploitation. However, it does not mean that the level of exploitation is not taken into account at all. Those who obviously want to fish for highly commercial purposes, such as for grouper, will not be allowed to do their business in the village's sea territory unless they sign a formal contract with the village regardless of how good their relationship is with the Dullah Laut villagers or even the village head himself.

Conclusion

Although this chapter discusses land and sea territory issues and illustrates their practical indivisibility, the main purpose was to introduce the basic features of traditional marine tenure in practice that will become the main focus of discussion in the following chapters. In conclusion, I will emphasise some points in relation to traditional marine tenure in Dullah Laut.

The sea territory under the ownership of Dullah Laut Village covers the entire waters surrounding the nine islands of Dullah Laut Village land territory. People

---

6   This norm does not apply to a territory under *sasi* regulation.

do not divide the sea territory into divisions as is done in Watlaar, but they do identify their territory by names that are given to specific places within it. If we compare the detailed division in Watlaar with the lack of division in Dullah Laut, we note that concerns related to boundaries are also different. In Watlaar, the boundary of their territory is defined by land boundaries with their neighbouring villages and by the outer division of their sea territory. Dullah Laut Village does not have land boundaries with other villages since its territory is archipelagic. They also do not define their sea boundary as is done in Watlaar, but they claim that particular coastal areas are theirs. For example, when they rejected the claim of ownership over Metan Er by Ut villagers, Dullah Laut villagers argued that the particular fishing spot called Metan Er was theirs. They did not include an explanation of the boundaries of their sea territory and show that the disputed spot was located within the demarcated areas, which may have been done if the case had happended in Watlaar territory. The different approaches to territorial boundaries between Watlaar and Dullah Laut may suggest that traditional concepts of sea boundaries in the Kei Islands are relatively flexible.

Another important feature of traditional marine tenure in Dullah Laut is that they differentiate between two different rights over the sea territory. The use right covers all activities related to extracting the sea resources such as fishing and formerly coral mining.[7] The property right comprises more rights than use right because it allows use rights as well as the right to transfer those use rights.

In the Kei Islands, the existence of the two rights is very important in relation to discussions of communal tenure. Discourse on communal property rights often makes the assumption that there is only a single right attached to the sea: either a right of ownership or a territorial use right (Christy 1982; Pollnac 1984 for the latter). Researchers investigating communal tenure in the region were not aware of the existence of different types of rights such as those in the Kei Islands and as such, did not analyse how each right was distributed within the community. Based on this false premise, inaccurate generalisations have been made such as: '[w]ithin the community the rights to the resource are unlikely to be either exclusive or transferable; they are often rights of equal access and use' (Feeny et al. 1990: 4).

Looking at how use rights and property rights are distributed within Dullah Laut, I would argue that communal marine tenure operates differently. It is correct to say that the use right is distributed equally to all members of the community. In fact, those who have an affinal relationship with community members might share in the right. However, the property right is controlled

7   People used to use coral for house construction before cement become more available. The Indonesian government now prohibits coral mining.

exclusively by the orgin *fam*, according to their version of local history. On the other hand, an alternative version proposed by Mr A. Rahaded, suggests that the descendants of the first village head are the exclusive right holders. Thus, the assumption that communal marine tenure means that all community members share equal rights to the territory or resources is not appropriate. In addition, the fact that the origin *fam* or the village head or Mr A. Rahaded signed contracts with various outside agencies proved that in practice, the communal use right over the sea territory is transferable. Therefore, the assumption that the communal property right is not transferable is not correct. This assumption applies to the property right but not use right.

In closing this chapter, I would like to emphasise that besides the flexibility of the concept of boundaries and the transferability of the rights, claims over sea territory are also based on narratives of origin which are subject to multiple interpretations and multiple versions (see Chapter Four). Therefore at a practical level, the practice of traditional marine tenure might become a source of conflict rather than a basis for developing a sustainable and socially just system of marine resource management. Various aspects of such conflict are explored in the following four chapters.

6. The Village Politics of Marine Tenure: Raiding 'Illegals' in Dullah Laut[1]

A day before I left Dullah Laut, I was involved in a raid of an 'illegal' fishing company that was preparing for a fishing operation in the Dullah Laut sea territory.[2] This incident developed into a serious conflict involving not only villagers and the fishing company but also military officers. At the village level, the conflict appeared to create three opposing factions. The involvement of military officers added to the complexity of the conflict because they represented outsiders' interest in the problem.

Although, the nature of this conflict was quite complex, encompassing many issues within the community and also relating to the outside world, I suggest that the political circumstances in the village were most influential in colouring the conflict. What I mean by political circumstances refers to the long-standing contestation between the descendants of the traditional village leader (*orang kaya*), the modern village leader (*kepala desa*), and the Christian settlement leader (*bapak soa*) over the position of village leadership. In this regard, these three political leaders used the customary marine tenure as 'political capital' to win the contestation. For the traditional leader, his control over village sea territory was used to gain economic and political support from the fishing company and military officers for his move to oppose the modern village leader. The modern village leader saw this as a challenge by the traditional leader to his position as formal leader of the village. So, raiding the 'illegal fishing company' was seen as his duty to restore his leadership in the village. For the Christian settlement leader, settling the problem that was triggered by the incident was a golden opportunity to show his own leadership abilities, something he had dreamed of doing for a long time. Thus for these leaders, controlling traditional marine tenure was a matter of being the village head. This meant that the political value of the marine tenure was much more important.

---

1   A shorter version of this chapter was published in MAST (Adhuri 2004).
2   After this incident, I stayed in Tual for a week before I left the Kei Islands for Jakarta. However, my interest in understanding this incident forced me to go back to Dullah Laut Village to attend a customary meeting and record some interviews. Although I did not go back to the village, I was able to understand what has happened after the customary meeting through speaking to some of the villagers who went to Tual to sell their fish and other business.

The Incident and its Resolution

Raiding an 'Illegal' Fishing Company

I was in my bedroom when I heard some people talking loudly in the guestroom. When I came out and asked them what was happening, they told me that they were discussing their plan to raid an 'illegal' fishing company on Rumadan Island. They talked about seeing three speedboats that had passed the village several times over the last few days laden with goods. At first, the speedboats were loaded with construction materials such as planks, timber and roofing material made of sago leaves. People wondered what they were going to do with those things in their territory. Later on, they saw the boats transporting fishing equipment, such as air compressors, hoses, nets, floating devices (empty plastic drums), as well as oil and fuel drums. Because such equipment is predominantly used by grouper fishermen, the villagers were suspicious of the outsiders' plans. Also considerable concern was mounting because these practices are often associated with the use of cyanide which is used to stun fish, making them easier to catch. It was when people saw a satellite dish, a television and an electric generator on board one of the speedboats that they were prompted to plan a raid on the fishing company. From this evidence, they believed an illegal grouper fishing company was ready to exploit and pollute their sea territory.

Having reached this conclusion, preparations were made to raid the company that very day. It was around two-thirty in the afternoon when they were ready to leave. A descendant of a well known war commander stepped on board followed by about twenty others. I noticed that leaders of three different *fam* joined this team. Some were villagers who worked outside Kei but mostly they were well-educated civil servants who were taking a holiday for an Islamic festival after Ramadan.

On the way to Rumadan Island we collected some people from the Christian settlement. The most important person was the leader of the settlement who was also the acting village head since the real village head was away. Other than the settlement leader, there were not many people from the Christian settlement who joined in. We left Ohoisaran with only the settlement leader and some youths and children.

Within fifteen minutes we saw the company's base camp. We observed two speedboats and a canoe equipped with an outboard engine anchored in the coastal waters of Rumadan Island. Some metres out to sea, we noticed two fish cages floating. On shore, we saw a house-like construction, half of which was completed while the other half was still under construction. In front of the house we observed air hose and rope (estimated to be more than 100 metres long) and some fresh vegetables. We also saw two air compressors, satellite dish, electric generator, another speedboat and an assortment of metal and plastic drums.

We were received upon arrival by a surprised Taiwanese man, his wife (a Javanese woman) and some workers (who were all Dullah Laut villagers). The 'raiding party' confronted the Taiwanese man and his wife because they believed that they must be the owners of the company. First, a man from the Muslim settlement asked if the company had a licence. They answered that they were arranging it with the descendant of the traditional village leader (Mr A. Rahaded) who they believed to be the owner of the territory. The man from the Muslim settlement then explained that the Dullah Laut territory was not owned by a single person, but by seven origin *fam* and that other *fam* living in Dullah Laut shared the use right of the territory. Thus, it was a mistake to arrange the licence with only one person. Another person—a university graduate— noted that although every Indonesian has the right to fish or establish a business anywhere in Indonesia, the law obliges them to observe certain procedures. For example, they should arrange a license with government offices starting from the highest down to the lowest levels. In this sense, the village government was the lowest authority with which the company should have arranged the licence. To this explanation, the war commander descendant added that even if the traditional village leader agreed, this agreement could not be valid unless it was approved by the seven origin *fam*—the owners of the territory.

The leader of the Henan *fam* who worked at the Southeastern Maluku Court office in Tual, continued the interrogation. First, he asked for the names of the Taiwanese man and the fishing company. The man's wife answered these questions. She said that his name was Mr C and after some thought she said that the company's name was CV TT. She explained that Mr C had nothing to do with this activity. 'He is only visiting me, his wife', she said. She also explained that she was the owner of the company and not her husband. Questions were then directed to the Javanese woman as the director of the company. Mr Henan took note of the information given by her. Finally, he said, 'Okay, we will process this in accordance with the law'.

The raiding party then demanded that the company stop their activities, take their belongings back to the capital city of Southeastern Maluku Regency, and wait until the 'real' (modern) village head was back to discuss the matter. The company was given one day to get off the island and they were warned that if they did not do this by the next afternoon, no one would take responsibility if people from the village took matters into their own hands.

After having been bombarded by questions and explanations, Mr C seemed to be confused. This was not only because his Indonesian language comprehension was not good but also, I believe, he thought that he had followed the correct procedures as suggested to him by the descendant of the traditional village leader, Mr A. Rahaded. In this regard, his wife explained that she had been arranging the licence. In fact, she had even sent a draft of the agreement[3] to the Christian settlement leader.[4] 'Additionally', she said, 'my company has not started doing any business. We are only making preparations. Thus, we have not made any mistake'.

It seemed to those present that Mr and Mrs C's arguments were a stalling tactic to buy them some time. They sent one of their workers to Dullah Laut Village to report the incident to Mr A. Rahaded. They wanted to avoid negotiations before Mr A. Rahaded arrived, but it was impossible not to respond to Mr and Mrs C while waiting for Mr A. Rahaded.

The negotiations started when their worker returned with one of Mr A. Rahaded's sons, who told Mr C and his wife to agree with what the people demanded. When Mr C and his wife refused, Mr A. Rahaded's son ushered them into the house to discuss the issues. After some time, they called the settlement leader to the house but as he came with other villagers, Mr C and his wife refused to negotiate. When they came out of the house, Mr C agreed to stop his activities and go back to the capital city of the regency. He asked the people to give him a day to pack and leave, to which they agreed. In return, the people asked the company to surrender one of their speedboats which would be returned when the company had done what they had agreed. A written agreement was prepared, read and signed by the settlement leader and Mr C's wife. That brought the raiding incident to a close.

The Customary Meeting

The signed agreement tendered by the company appears only to have been a strategy to calm the people who raided them in Rumadan because the night of the incident, Mr C, his wife and Mr A. Rahaded, accompanied by a soldier from the local army post, forced the settlement leader to hold a customary meeting to discuss the possibility of granting the licence so that the company could continue its activities. The settlement leader had no power to refuse so he called together some elders in his settlement and brought them all to the village head's house in the Muslim settlement.

---

3   When people talked about a licence, it usually refers to an agreement between the modern village leader and other leaders, and the company's representative. The agreement states that the village grants the company the right to operate in their territory. In return the company gives an amount of money to the village.
4   Later when I spoke with one of his workers I was told that she was right —the letter of agreement had been prepared. It was signed by Mr A. Rahaded, the local army post's commander, and a blank space was left for the modern village leader's signature.

They arrived at the village head's house about nine-thirty in the evening. The village head's father and some other elders in the Muslim settlement received them. After the settlement leader mentioned their intention, the village head's father and his supporters refused the proposal on the grounds that the modern village leader was away. They also referred to the agreement signed by the company and Christian settlement leader earlier that day which ordered them to stop operations and return to the capital city. Although the company, backed up by military personnel and the settlement leader, kept insisting on holding the meeting, they were unsuccessful. The head of the village's father and his supporters sent the company on their way.

The situation had grown more intense the following day. The company and Mr A. Rahaded used the local army post commander to pressure the other parties. Early that morning, the Christian settlement leader was picked up from his house. He was taken to the office of the military post commander in Tual. According to the settlement leader, the military commander ordered him to pursue the case by holding a customary meeting to discuss the company's proposal that afternoon. The customary meeting had to reach a decision on whether or not the company would be allowed to operate in their territory. If the meeting decided to refuse the company's proposal, the military commander ordered that the settlement leader and other village leaders at the meeting prepare a letter stating that their village would not accept a similar proposal from any other company. In other words, if the company's proposal was refused, the village should declare that their territory was closed to outside fishing companies.

The modern village leader's supporters also reported the case to the local authorities. Early in the morning, they went to the local police station where, having already heard about the case, the station commander promised to call all parties involved to meet with him the following day. After that, they went to the army post to meet the military commander who of course already knew about the case because the company and Mr A. Rahaded had come to see him the day before. In fact, he had sent a letter 'inviting' the settlement leader to come to see him. He regretted that people had gone to the police post because it looked as though he was in opposition to the commander of the police post. He asked the people to meet him the following day together with the other parties involved.

I was surprised that afternoon when a villager came to my place in Tual. He told me that the situation in Dullah Laut was very tense. The settlement leader was arranging a customary meeting to decide the fate of the company. I was told that the settlement leader, Mr A. Rahaded, Mr C and his wife, and a soldier had been waiting in the village since midday. The villager had just informed some *fam* representatives who were still in Tual after reporting the case to authorities and was on his way back to Dullah Laut. I was offered a lift and asked if I wanted to observe the meeting. Of course, I was very keen given the emotionally charged state of the villagers and because this meeting was very much relevant to my research.

The settlement leader's guestroom was full of people when we arrived around five in the evening. About 15 people were already inside. They were the settlement leader, a soldier from the local military post, Mr C and his wife, Mr A. Rahaded, and representatives of origin *fam* and the Christian and Muslim settlements. This was a very special customary meeting, not only for me but also for all the villagers because this was the only meeting where the origin *fam* from the Muslim settlement was represented by two groups of *fam* leaders, indicating the political factions in the Muslim settlement.

The meeting was opened by the settlement leader who made three points in his opening remarks. First, he noted that he had reported the meeting to the military commander earlier that day and explained the commander's wishes for the meeting's outcome. Then, the settlement leader criticised the absence of the village head. He said that the modern village leader should have been back by that time since he had asked permission to leave the village for only two weeks. In addition, the reason for his leaving was not official but personal, which meant that his trip was not for the benefit of the villagers but for himself. Third, for these two reasons he said it was his role to lead the meeting and decide whether the company would be granted permission.

A soldier representing the military commander gave the second speech. He explained he was there to ensure that the problem was handled peacefully. He asked the people to solve the problem at once so that further conflict could be prevented. Like the settlement leader, he also emphasised what his commander expected from the meeting. At the end of his speech he criticised those involved in the incident the day before. He said that the company had not begun its operations and there was not enough evidence to accuse them of using cyanide. Therefore, it was wrong to confiscate their speedboat. He also regretted that people had sworn at Mr C's wife and asked the people to return the speedboat.

When the settlement leader asked the people to express their opinion, they started the discussion by answering the soldier's remarks. Mr T. Nuhuyanan, the owner of the boat used to raid the company, explained that the speedboat was not confiscated. It was surrendered voluntarily as a guarantee that the company would leave Rumadan Island as stated in the agreement signed by the company representative and settlement leader. The speedboat would be returned when the company left the island. A man from the Muslim settlement took up this point. He provided the legal definition of the word 'confiscation' and said that the incident did not fall under this definition. He also explained that it was the right of the people to defend their territory from outside intrusion. He also asked why this incident was being questioned while the 'illegal' presence of the company was not considered to be a problem. Regarding the company's operations, he questioned their use of an air compressor with such long hoses. He suggested that it was illogical that the company would only use fish traps and line fishing—as the company had told them—with such a compressor. These ideas were supported by other representatives who also expressed some additional concerns such as ecological destruction, the economic impact, and the fact that the company's presence had driven people to fight with each other.

Some representatives questioned these arguments. A man from the traditional village leader's faction raised the issue of representation. He said that the village was divided into three political factions and those who raided the company did not represent all three factions. He added that those who raided the company were youth who worked outside the village and their representation of both the origin *fam* and the village itself was questionable. He also raised the issue that surveillance of company activities was not their responsibility and required government officials. The fact that the company had been granted a licence was proof that its activities were legal. He argued that even if the company had abused its licence, the people had no right to punish them.

Mr A. Rahaded (the traditional village leader's descendant), then took up the discussion. He explained that the company had not come without permission. It was he who had allowed them to operate in Dullah Laut territory and construct their base camp on Rumadan Island. He added that the settlement leader had been notified of the plan about two weeks before the raid. At that time, Mr A. Rahaded had told him that a fishing company might come and fish in village territory. 'Now, since the incident has happened, let us stop accusing them of being "illegal", because even if it was wrong, it was my fault not theirs', Mr A. Rahaded said. 'Now, let us hear the opinion of all representatives as to whether we will grant them the permission [to have a base camp and fish in village territory]. I would like to hear from each of you.'

The settlement leader took up this point and tried to continue the discussion. He started by saying that the company's representatives were surprised and scared so they agreed to sign the statement. According to the settlement leader, what they really wanted was to be allowed to pursue their activities. 'Thus, they now come to us to propose their intention.' He then asked each of the representatives to express their opinion. A representative from the Christian settlement agreed and asked other representatives from his settlement to actively participate in the discussion.

The discussion however, did not go in the direction that Mr A. Rahaded and the settlement leader wanted. Another representative from the Muslim settlement, who was involved in the raid, interrupted the discussions by evaluating the authority of the meeting. He did this by criticising Mr A. Rahaded, who had given the company permission as if he had the authority to transfer the ownership or use right of their sea territory. 'This was not right', he said. He also verbally attacked the settlement leader for washing his hands of the incident the day before and concluded that this was a sign that the settlement leader was inconsistent. He demanded that they postpone the discussion until the real village head was back. 'We are now walking without the "head", we will only be complete as a human being when the real village head is back here.' This idea was supported by the descendant of the war commander and others who had paid more attention to raiding the company than pursuing the discussion about granting the company's licence.

It was almost eight in the evening. The sun had set, forcing us to turn on a gas lamp in the house. The meeting progressed slowly. It was primarily an argument between those who wanted to pursue giving the company a green light to continue their business and those who wanted the company to leave and wait until the modern village leader was back.

Finally the soldier took control of the discussion. He did not see any way the meeting would reach a conclusion and considered that the meeting had caused considerable conflict rather than reaching a solution to the incident. Therefore, he stopped the customary meeting without any agreement being reached.

Political Autonomy of the Settlement and Modern Village

To understand the various undercurrents of conflict in this incident, an analysis of the different forms of political autonomy in Dullah Laut is required.

The Settlement

Villagers believe that by tradition, Dullah Laut is an autonomous village. This means that they have the right to govern themselves. There are two primary characteristics associated with self-governance. The first aspect concerns social relations—people believe that they have full authority to control all social relations in the village. This is what I call 'social autonomy'. The second aspect concerns issues related to territory. People believe that as a social unit, they control their own territory meaning they believe that they have the right to distribute and make use of their own territory. This is what I call 'territorial autonomy'.

In terms of social autonomy, Dullah Laut is divided into two settlements. The two hamlets—Ohoislam and Ohoisaran—physically represent this division. According to the narrative of origin, this division was created with religious conversions. Those who converted to Islam around the 1850s moved from the original settlement and erected a new hamlet on the eastern tip of the island. Their settlement was what the Dutch sources called 'Tewaniohoi' (see Riedel 1886: 222[5]), which became Ohoislam when the villagers moved to the current settlement. The original settlement populated by those who converted to Catholicism in the late nineteenth or early twentieth century was called Duroa by the Dutch sources and is now called Ohoisaran.

In terms of territorial autonomy, Dullah Laut as a 'traditional village' is not divided into smaller units. So even though the village is socially divided into two settlements, both settlements share an undivided territory. In Dullah Laut, people don't refer to *petuanan* Ohoislam or *petuanan* Ohoisaran when they discuss issues of territory. They use the term *petuanan* Dullah Laut.

Let's now discuss the internal structure and distribution of power in a village. According to Geurtjens (quoted in Van Wouden 1968: 36–7) there were five prominent functionaries in a village. These were: the traditional village leader (*orang kaya*); lord of the land (*tuan tan*); the attendants of the local spirit (*mitu duan*); Islamic religious official (*leb*); and precursor and carver (*dir-u ham-wang*). The traditional village leader was the headman of the village who 'used to be a particularly independent governor in his village' (Van Wouden 1968: 36). On this particular point, Van Wouden comments:

> most probably we should take this to mean that each village formed a practically independent unit, for in fact any tendency towards such independent rule was entirely alien to the office of headman. He was not permitted any arbitrary action, and for all important questions he had to call a meeting of the "elders" of the family groups (ibid.).

The lord of the land was 'the official owner of all village lands'. In times when the lord of the land still held authority, his role was crucial in allocating their territory. He was the person villagers would go to whenever they wished to make a new garden. He was also considered to be the person who knew most about land distribution among people in the village and boundaries between neighbouring villages. Therefore, he played an important role in solving disputes over land ownership.

The attendant of the local spirit was responsible for dealing with affairs related to the ancestral spirits and local guardian spirits, and the religious leader was responsible for the Islamic rituals. Their role was to perform sacrifice rituals on behalf of the community. Finally, the precursor and carver handled matters related to the ceremonial war boat (*belang*), the emblem of the village. He piloted

---

5 Riedel spells it Tawaniohuit.

the ceremonial boat during its departure from and arrival to the island and was also the person responsible for distributing the catch of communal hunting or fishing to the villagers.

It is quite difficult to understand the exact power structure of these village functionaries since Geurtjens did not provide details explaining the relationships between those who share power. It is clear though that the political power of the village elders was superior to that of the traditional village leader and other functionaries and held the spot at the top of the organisational structure (see Figure 6-1). The superior position of the elders allowed them to give direct commands to each type of functionary (shown by bold lines). The village functionaries are not represented at the same level as the traditional village leader (*orang kaya*) because while these functionaries might represent the totality of village political power, each of them holds only a specific power. For example, during a dispute over land or other problems related to territorial autonomy, the lord of the land (*tuan tan*) might play a leading role. In another context—such as marriage— the attendant of the local spirit (*leb*) and religious official might be centre stage because this is the context in which their respective powers are required. In this regard, we might say that there is no permanent hierarchical relationship between the village functionaries. However, since the traditional village leader was the governor of the village, it seems that in every situation his role was needed.

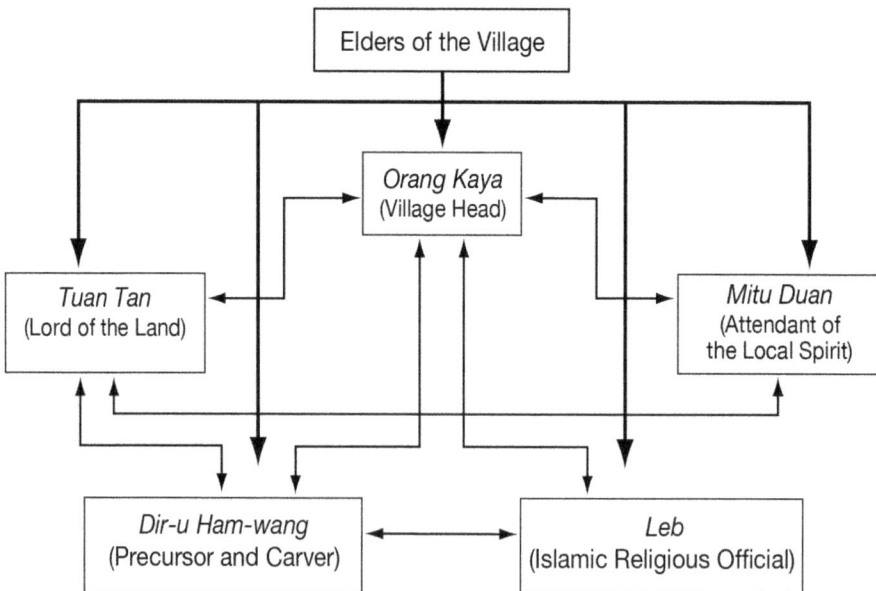

Figure 6-1: Structure and distribution of power in a traditional village.

Source: Author's fieldwork.

Theoretically, the political life in Dullah Laut is in the hands of a committee comprising members of the origin *fam*, *Ohoiroa Fauur* (see Chapter Four). In the political realm of the village, this committee is considered to be the holder of ultimate power. This means that they control all issues relating to Dullah Laut as a village, both in social and territorial terms. This control works in both inward and outward directions, meaning that the *Ohoiroa Fauur* has the power to control the whole population of Dullah Laut in matters relating to the social order of the village, and represents the interests of the village to the outside world. For example, if there is a conflict between villagers that the smaller social group cannot solve, the *Ohoiroa Fauur* will hold a meeting with all parties involved in the conflict. In this meeting the *Ohoiroa Fauur* will examine the nature of the conflict, decide which party is at fault, and find a solution to the conflict. Once the *Ohoiroa Fauur* has reached a decision, all parties are required to comply with the outcomes. The *Ohoiroa Fauur* also represents the interests of the village to the outside world. So for example, if a fishing company wishes to operate in Dullah Laut territory, it is the *Ohoiroa Fauur* from whom the company should get permission.

In practice, the *Ohoiroa Fauur* distributed their power to what might be called the village functionaries. As a settlement, Dullah Laut had only five village functionaries. They were the traditional village leader, war commander, Muslim leader, and the two settlement leaders. In theory, the traditional village head was the governor of the village. The war commander was responsible for handling potential or real conflicts, particularly with outsiders. For example, during wartime it was his duty to devise the war strategy and to coordinate villagers' roles within it. The imam was responsible for dealing with issues related to Muslim religion and rituals. Conceptually, in terms of social autonomy, the settlement leader was also important because the settlement organization was under his leadership.

Among the five functionaries, the traditional village leader was the most important mainly because there were so few functionaries in the village. This caused the political power of *Ohoiroa Fauur* to be distributed amongst a limited number of people. If we consider the non-existence of the lord of the land[6] alone, this resulted in the traditional village leader becoming the power holder regarding both social and territorial autonomy. Other factors that have caused the traditional village leader to become so prominent include the diminished role of the war commander since conflicts that generated what used to be called war rarely occur now. In addition, the role of the Muslim leader has diminished because there are few communal rituals performed in the village and because

---

6 The explanation for the extinction of the lord of the land can be found in Chapter Four.

the imam only serves the Muslim community which is isolated from the Catholic population at the Christian settlement. The third factor is that the leaders of both settlements are under the direct command of the traditional village leader.

To understand the third factor requires an analysis of the internal structure of the Muslim and Christian settlements. I mentioned earlier that in terms of social autonomy, Dullah Laut was divided into two settlements and for issues regarding the internal life of the settlement, they were run autonomously. For this purpose, each settlement had its own *Ohoiroa Fauur* representatives consisting of the head of the origin *fam* living in each settlement.[7] As at the village level, they were considered the political power holder in the settlement.

In turn, the *Ohoiroa Fauur* of each settlement delegated their power to the settlement leader. In theory, the *Ohoiroa Faur* of the Christian settlement appointed the settlement leader to exercise their power in maintaining harmonius relations in the settlement. Likewise, the *Ohoiroa Fauur* of the Muslim settlement provided the settlement leader with the power to govern the settlement. By this delegation of power, the control of daily life of both settlements was in the hands of the settlement leaders. In the Muslim settlement however, the practical leadership was directly in the hands of the traditional village leader. It was only in the Christian settlement that the settlement leader exercised some level of autonomy. During my fieldwork, I observed that only in the Christian settlement did the settlement leader hold a customary meeting. The settlement leader consulted the traditional village leader when he faced a problem he could not handle himself. In this context, he would follow the decision of the traditional village leader. Similarly, customary meetings at the Muslim settlement were always led by the traditional village leader.

Finally, having explained the structure and the distribution of power in the traditional village, I would suggest that the political power in Dullah Laut was not distributed in the way suggested by Geurtjens. The political power in Dullah Laut was more centralised in the hands of the traditional village leader as shown in Figure 6-2. This is not only because of the reasons outlined, but also because the village functionaries are taken to be representatives of the origin *fam* (*Ohoiroa Fauur*). The imam is from Nuhuyanan, the war commander is from Rahawarin, and the settlement leaders of the Muslim and the Christian settlements are from Raharusun and Rahawarin respectively. Thus, since the role of the war commander and imam has decreased in importance and the heads of

---

7   In fact the *Ohoiroa Fauur* of the village was a coalition of the *Ohoiroa Faur* of Ohoisaran and *Ohoiroa Fauur* of Ohoislam. It was not uncommon during a customary meeting at the village level for each *Ohoiroa Fauur* to represent the interests of their own settlement.

both settlements are under the traditional village leader's command, the role of the traditional village leader has increased significantly in importance and even has some control over *Ohoiroa Fauur*.

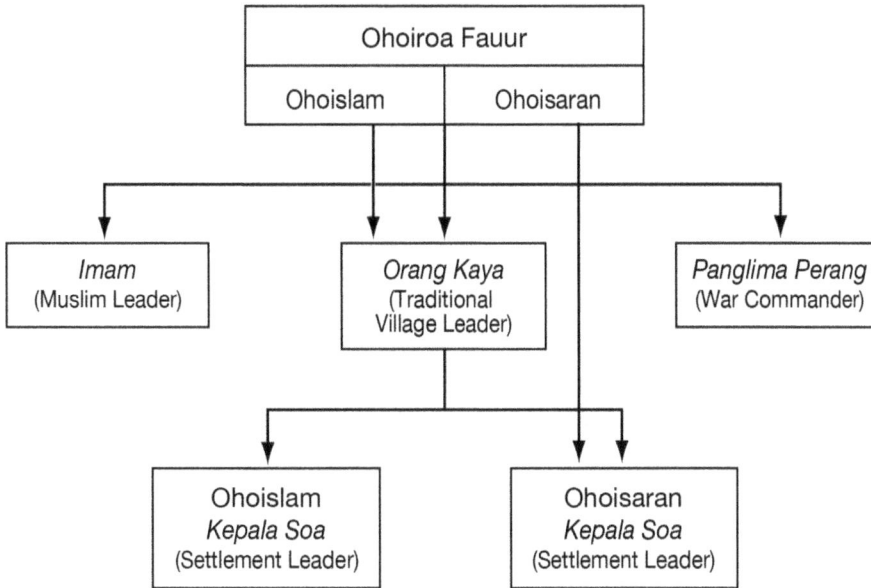

Figure 6-2: Traditional structure and distribution of power in Dullah Laut.

Source: Fieldwork research.

The Modern Village Organisation

Dullah Laut had been regarded as consisting of two different villages (*desa*[8]) since the Kei Islands became part of the Republic of Indonesia in the early 1950s up until 1989. Formerly, the Muslim and Christian settlements were called Desa Dullah Laut Islam and Desa Dullah Laut Roma Katolik (RK) respectively. Being considered two distinct villages meant that Dullah Laut Islam and Dullah Laut RK were autonomous units with full rights to govern their own people and territory. Each village was led by a different leader who had their own staff and village deliberation council. Thus, when the central government started providing village subsidies in the 1970s, each village received the same amount and was independent of the other's influence in making use of the subsidy. Of

---

8   The legal terms used for a village, its leader, staff and its legislative assembly have changed over time with changes to the village law. For example, in *Village Law No. 19, 1965*, the *desa* was called *desapraja* and the title of the head of the village and his staff depended on local tradition. In the *Village Law No. 5, 1979*, the village is called *desa* or *kelurahan* and staff are named differently according to their position, for example the village secretary is called *sekretaris desa* and the development program coordinator is called *kepala urusan* (*kaur*) *pembangunan*.

course overall they were both bound by the same obligations under Indonesian law to report the allocation of their subsidy to the head of the subdistrict office in Tual.

Despite the fact that the separation of Dullah Laut was not in accordance with tradition, few problems arose between the two villages. In fact, there were many factors that created relatively harmonious relations between Dullah Laut Islam and Dullah Laut RK. The first factor was that the villages did not have much exposure to the external world and the village government mostly dealt with only internal issues. Since both the Muslim and Christian settlements were relatively autonomous, this caused few reasons for interaction or subsequent conflict between the two villages. Secondly, whenever they dealt with outside agencies—particularly regarding the use of their territory which according to tradition is inseparable— each head of the village was allowed to represent the interest of both villages, and when a large amount of money was involved, they would make a decision together and share the risks and the benefits. The latter implies that the Indonesian village law, which considers every village to have its own territory and make its own decisions indvidually, was not strictly followed to the letter.[9]

Furthermore, some have suggested that the separation was beneficial citing for example, that as two different villages they got two packets of central government subsidies allowing the two to develop their villages better. Another example of the benefits of the separation occurred when the Madurese fishermen came to ask permission to fish in their territory. Both heads of the villages asked for the 'betel nut money' (the customary term used for a contractual fee) from the fishermen. They believed that they could not have done so if they had been considered a single village.

Nevertheless, the Indonesian government's equal treatment of the two villages promulgated a profound change in the traditional relationship between these two settlements. Some informants told me that during this period the village head of Dullah Laut RK often brought problems in his village directly to the King of Dullah meaning that he ignored the role of the traditional village leader who sat in Dullah Laut Islam. According to tradition, the head of the Christian settlement should have brought his problem to the traditional village leader. Only if the traditional village leader could not solve it, should the problem have been taken to the king. In this case, the traditional village leader would lead

---

9   All of the village laws, including the Dutch Village Law which was applicable until the Indonesian government substituted it with the Village Law No. 19, 1965, implicitly or explicitly considered that a village must have its own territory (see Marsono 1980).

the settlement leader and those involved in the issue to meet the king. What is implied by this example is that the hierarchical relationship defined by tradition between the Muslim and Christian settlements was starting to lose relevance.

In 1989, the governor of Maluku issued a *Provincial Decree No. 146/SK/39/89* which regulated the number and names of *desa* and *kelurahan*[10] in Maluku Province. According to this decree, Dullah Laut was considered to be a single village that bore the same name. This decree led to the understanding that the village of Dullah Laut consisted of two hamlets (*dusun*), Dusun Dullah Laut Islam and Dusun Dullah RK. Since the seat of the modern village leader was at Dullah Laut Islam, it meant that Dullah Laut Islam was the centre of the village (*desa induk*, lit. mother village) and Dullah Laut RK became the 'child village' (*anak desa*).

This decree was a surprise to the people of Dullah Laut RK, particularly the former village head. It was a surprise because in September 1987 he and the village head of Dullah Laut Islam had arranged a meeting attended by representatives of origin *fam* (*Ohoiroa Faur*) from both villages. The aim of this meeting was to change the name of each village since they shared the same words—Dullah Laut—and the words Islam and RK represented a religious division that might have bad connotations. The meeting decided that the names Dullah Laut Islam and Dullah Laut RK would be changed to Dullah Laut and Duroa respectively.[11] It was clear there was no indication at this meeting that the two villages would be merged. In fact, the aim of the meeting was understood to have been an attempt to strengthen the division between the villages.

The decree was therefore inconsistent with the direction taken by the former village leader of Dullah Laut RK who had tried to loosen the hierarchical relationship with the Muslim settlement and establish the notion that the Christian settlement was independent from the traditional village leader at the Muslim settlement. By contrast, the 1989 decree degraded the position of the Christian settlement from an independent village to a 'child village' (*anak desa*), which meant it was under the control of the modern village leader at the Muslim settlement.

The people of the Christian settlement, or at least the former village head, questioned the decree. They were suspicious that some people had misused the letter they'd signed as a result of a September meeting a year before which was an agreement that Dullah Laut RK would become the '*anak desa*' while Dullah

---

10 *Kelurahan* is another term used for village but unlike *desa*, a *kelurahan* has no autonomous right in any sense. The *kelurahan* leader and programmes were appointed and arranged directly by the Indonesian government. Villages in a city are governed as *keluarahan* (see Village Government Law No. 5, 1979).

11 A report of the result of this meeting was made and sent to the head of regency Maluku Tenggara, signed by both village leaders. The list of those present at the meeting and their signatures were attached to the letter.

Laut Islam would be the centre of the village should both villages be merged into a single village. They also argued that if the village was to be considered a single *desa*, it was the Christian settlement that should have been the centre of the village (*desa induk*—mother of the village). Interestingly, the latter argument was developed based on another interpretation of tradition. They believed that the decision on the location of the centre of the village and child village should have been based on the origin of the settlement and not on the seat of the village head. In this sense, the Christian settlement should have been the centre of the village because the village centre or 'navel' (*woma*) was located in this settlement. This interpretation of tradition must be quite recent since the ruler of Baldu (Dullah) told me that it was he who had located the centre in the early eighties. At that time, he attended a ceremony for the construction of the church that was led by the first priest from the settlement. During a break, the priest asked him, 'If this is a village where is its *woma*?' At that moment, the ruler of Baldu spontaneously designated a plot of land at the corner of the football field located in the middle of the settlement and said: 'That is the *woma* of this village, with the name of *woma* Varne Harmas'.

I also spoke with others who argued that even when considering the position of the traditional village leader, the Christian settlement should have been the centre of the village. They explained that the first traditional village leader, Yahaw Rahaded, had lived and was buried at the Christian settlement. The position of traditional village leader was transferred to the Rahaded at the Muslim settlement when Yahaw Rahaded passed away because his oldest son, who was Catholic, worked outside of the village. This meant that his younger brother inherited the title and took the position. Thus, 'If we will follow *adat* correctly' they said, 'the title of traditional village leader should be brought back to the Christian settlement by appointing a descendant of the eldest son of Yahaw Rahaded to take the position'.

The former leader of the Christian settlment always raised this issue when he met officials whom he believed had the power to reconsider this issue. For example, when he was visited by a team representing the ruling party during the New Order of Indonesia (Golkar), he asked the leader of the team if Golkar could help raise the status of his settlement to a village as it used to be. The subdistrict leader of Kei Kecil and his staff also told me that the former Christian village head had raised the same question with them. Interestingly, the issue of turning their settlement into the central village—which meant reversing their relationship with the Muslim settlement—was never a concern of the former Christian settlement's leader. In an interview, he told me that according to Village Government law, a village head should be elected by the villagers. Considering that a higher proportion of the population is Muslim, it would be difficult to have a village head from the Christian settlement since it would be almost impossible for the Muslims to vote for a Catholic village head. Further, promoting the argument that the position of village head should have been at the Christian settlement because the hamlet is the origin settlement of the village would not be beneficial to the former Christian settlement's leader. The consequence of this argument would be that the position of village leader should be given to the descendant of the first traditional village leader, which the former village head is not.

Internal Structure

There are some laws that have become the basic reference for the village organisation. The first law was the *Inlandsche Gemeente Ordonnantie Buitengewesten*, the Dutch law that regulated the village organisation in the outer islands of the Netherlands Indies. This law was not intended to change the structure of the traditional village organisation but was used to benefit the Dutch, both politically and economically (Cooley 1973). However, in practice this law brought about many changes, at least in some villages. For example, the Dutch were involved in the appointment of the traditional village leader, and their appointments were not always in accordance with tradition. Another more important example was that the law degraded the position of the lord of the land because it did not differentiate between social and territorial issues. Under this law both issues were under the control of the traditional village leader. In some villages this change generated disasterous conflicts (see Chapter Nine).

Dullah Laut did not experience the 'negative' impact of the Dutch law. In fact, the Dutch period of village organisation is seen as the time when tradition was followed properly regarding the appointment of the first traditional village leader. People accepted Yahaw Rahaded as the first village leader. By doing this, they believed that the position of the traditional village leader was the right of his descendants. In addition, people did not see that the Dutch law created any change in power between the traditional village leader, the origin *fam* (*Ohoiroa Faur*), and village functionaries.

The Dutch village law was replaced by *Village Government Law No. 19, 1965*, which was in turn replaced by *Village Government Law No. 5, 1979*. Many changes have been brought about with the application of these laws, especially the latter. Two of these changes are worth mentioning here. First, regarding the distribution of power in the village, I believe that the ideology of these laws is centralisation, and that the head of the village is the centre of all elements in the village organisation.

Both laws consider that the head of the village is what Geurtjens might call the governor of the village (art. 10 of the Village Law No. 5/79). Van Wouden's comment that the head of the village was not a real governor since he was expected to consult elders of the village was not applicable here. It is true that for important issues, the village leader should consult or be responsible to the Village Deliberation Council (*Lembaga Musyawarah Desa* or LMD) which is the representative body of the villagers. But the laws have also assigned the leadership of the LMD (art. 17/2 of Village Law No. 5/79) to the village head. The village head also controls his staff which consists of: a village secretary (*sekretaris desa*); an administrative coordinator (*kepala urusan, kaur pemerintahan*); a development program coordinator (*kaur pembangunan*); a welfare program coordinator (*kaur kesejahteraan*); a treasury coordinator (*kaur keuangan*); a general coordinator (*kaur umum*); and heads of constituent hamlets (*kepala dusun*). All of these positions mostly take orders from and work for the village head, except for the heads of hamlets. The head of hamlet is different because he is considered to be the representative of the village leader in the hamlet (art. 7/2 of the *Village Government Law No. 5/79*). This means that a certain amount of the village leader's power is transferred to him.

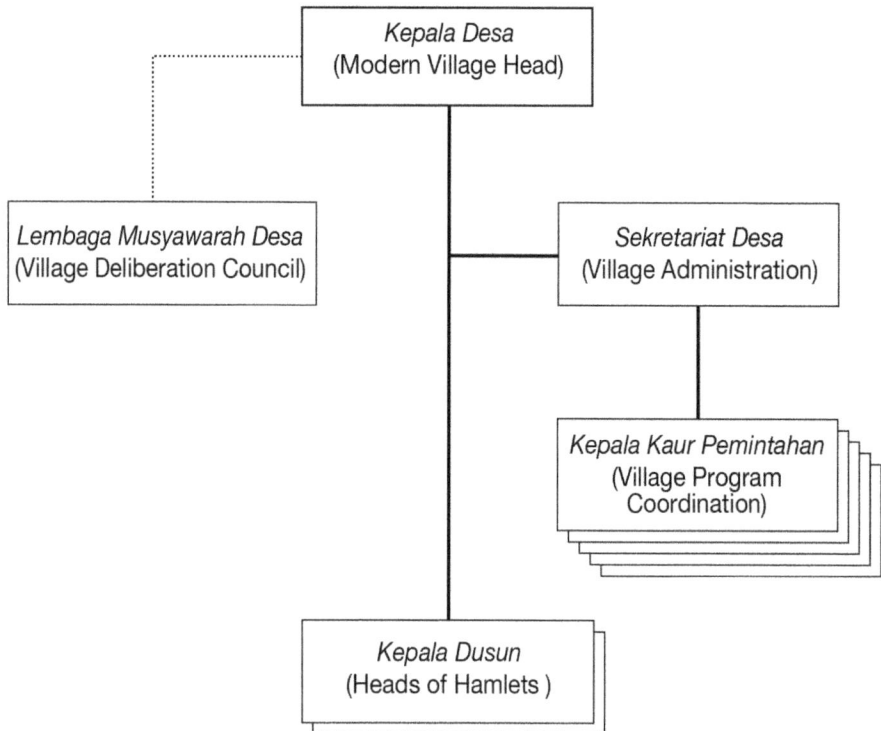

Figure 6-3: The modern village organisational structure.

Source: Adapted from Marsono (1980).

In practice, the role of the village head is obvious. He is the only representative of the village that interacts with the outside world. He is also the only person to whom the power of the central government is delegated. It is only through the village head that the Indonesian government provides subsidies and programs from diverse outside sources. All problems in the village that need to be resolved outside of the village should go through the village head before they are taken to the external agencies.

These laws also regulate the number of positions in the village organisation, who fills these positions, and the ways in which villagers can choose and assume such positions. While the *Village Law of 1965* still gave some acknowledgement to local tradition, the *Village Law of 1979* paid almost no attention to it at all. In fact, the main goal of the *Village Law of 1979* was the unification of the village organisation in Indonesia using a supposed Javanese cultural model (Kato 1989: 94).

It was predictable that the Kei people would not approach these issues in the same way. For example, under both laws the village head should be elected from and by the villagers. This violated the tradition which identified the position of traditional village leader (as well as other village functionaries) as inherited positions. These laws were also not in accordance with the tradition that considered the village leadership to be the privilege of the *mel*. According to the tradition, the *iri* had no right at all to take part in any matter regarding village leadership. However, the law stipulated that all villagers had the same right to vote and to be voted for in the election of a village head.

As noted previously, some villagers perceived that these laws did not replace tradition. They believe that the village laws are supplementary to tradition rather than contradictory. This understanding derives from some ambiguity of the law toward tradition. I mentioned earlier that in the *Village Law of 1965*, the titles of the village leader and functionaries still used local names. As noted by Kato (1989), in the 1979 law, the ambiguity can be found under point b) in the section headed 'to consider' (*menimbang*), which notes:

> In accordance with the nature of the Unitary State of the Republic of Indonesia, the state of affairs concerning Desa administration is to be made uniform as much as possible, with due respect for various local conditions of Desa and stipulations of customs (*adat istiadat*) still in existence, in order to strengthen Desa administration so that [we will be] more competent to mobilize society in its participation in development and to run Desa administration increasingly more extensively and efficiently (emphasis added, ibid.: 93).

Some other villagers considered that the village laws offered new options on how the village could be organised. These people acknowledged the difference between tradition and the law, but did not decide which to follow. In fact, they applied a mix of tradition and the law and explained that by saying 'as the children of *adat*, we cannot just leave the *adat,* and as citizens we should also follow the government'. As an example, the present village head said that it was difficult to make village staff appointments and had not yet done so because the number of staff stipulated by the law was not large enough to accommodate the number of persons appointed by tradition.

The ambiguous relations between Indonesian government laws and tradition and the various interpretations of how these laws should be applied generated problems in Dullah Laut. The biggest problem was the conflict between the descendants of the traditional and the modern village head at the Muslim settlement. The former claimed that the position of modern village head was his right while the latter believed that he was the one who had been chosen for the position by the villagers and approved by the Indonesian government. Another serious problem related to the position of the former modern village head at the Christian settlement whose position was degraded to that of hamlet leader and village secretary. The following section will discuss these issues in detail.

The History of Village Leadership

The main reference point for the discussion of village leadership tradition in Dullah Laut is the history of the traditional village leader. The starting point of the history was the appointment of the first traditional village leader in the last decades of the nineteenth century or the first decade of the twentieth century. According to this history, the first traditional village leader was Yahaw Rahaded, who was appointed by the Dutch. There was no conflict during his leadership and it seemed that the people of Dullah Laut welcomed his appointment. This appointment was the point of reference for the claim that the position of traditional leadership in Dullah Laut is the right of Yahaw Rahaded's descendants.

This belief was also the reason why people accepted the transfer of the position to Yahaw Rahaded's son, Yahaw Rahaded's son's son, and Yahaw Rahaded's son's son's brother-in-law respectively (Figure 6-4).

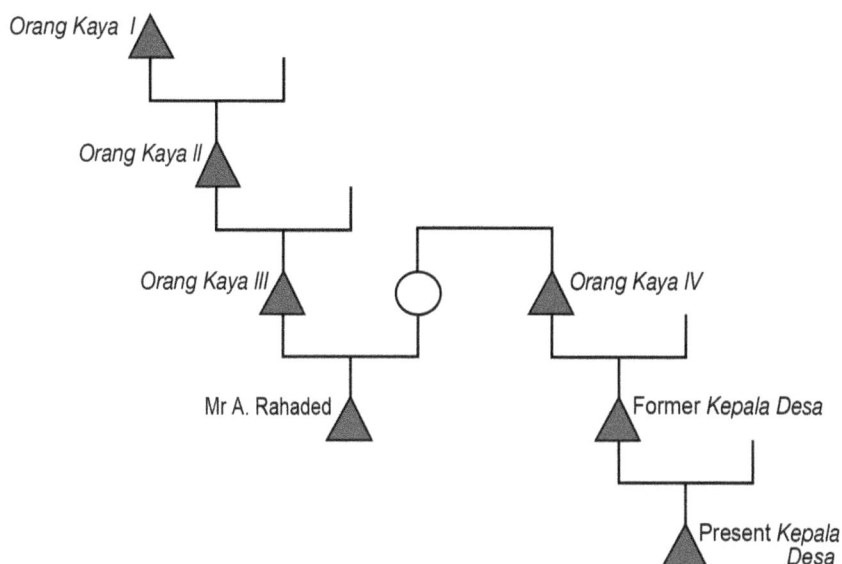

Figure 6-4: The genealogical connections of Dullah Laut leaders.

Source: Fieldwork research.

The succession of the first two followed the usual pattern for transfers of the title— from father to son. The transfer from the third traditional village leader was unusual because it was not from father to son but between in-laws. However, since the transfer was initiated by the legitimate person—the third traditional village leader, who was the direct descendant of the first—and for a legitimate reason—that there was no adult descendant of the traditional village leader—people accepted the leadership of the fourth traditional village leader. In such circumstances it was quite common for the title to be transferred to those who have close relations with the former title holder. In this context, an informant told me that during the leadership of the third traditional village leader, his brother-in-law Mr A.H. Nuhuyanan, often helped the traditional village leader in time of difficulties. This was the reason why the third traditional village leader transferred his title to Mr A.H. Nuhuyanan in the early-1930s.

One of the most significant signs of the peoples' acceptance of the fourth traditional village leader's leadership was the involvement of Mr A. Rahaded, a son of the third traditional village leader,[12] in the development of the village during the second half of the fourth traditional village leader's tenure. In 1963, Mr A. Rahaded initiated the establishment of an Islamic elementary school. This idea became a shared dream of all the Muslim villagers. They worked together

---

12   A. Rahaded was still a child when his father passed away. Therefore, the position of traditional village leader was not transferred to him but to A.H. Nuhuyanan.

in constructing the school. Some local volunteers were also prepared to teach in the school. During that period, their only mosque was also enlarged. Mr A. Rahaded donated his land for this purpose.

However, this development program was said to be the beginning of the conflict between the elite members of the village, particularly between Mr A. Rahaded and the fourth traditional village leader's *fam*. The conflict escalated when the connection between their village and the central government and outside agencies started to develop. This period began when the central government started providing a village subsidy in 1970 (Kato 1989) and when donations were received from some outside agencies for the construction of the elementary school. The conflict mostly involved the distribution of the money or materials they received.

The conflict worsened when the fourth traditional village leader died and his son, Mr M. Nuhuyanan, replaced him in 1967. Mr A. Rahaded disputed this transfer by claiming that the title should have been transferred to him. His claim was based on the fact that the three first traditional village leaders at Dullah Laut were his father's father's father, his father's father and his father. Explaining why his father transferred his title to Mr A.H. Nuhuyanan, Mr A. Rahaded argued that his father had entrusted him with the position because at that time, no one of his generation was old enough. According to Mr A. Rahaded, it was agreed that whenever the descendants of the first traditional village leader were ready to take over the position, Mr A.H. Nuhuyanan would resign and transfer the title back to them.

According to Mr M. Nuhuyanan, he did not return the position for several reasons. The first was because Mr A. Rahaded did not ask for his right 'politely,' and the second was that the Indonesian government recognised his position. The Southeastern Maluku Regency leader issued a letter for his appointment as the village head of Dullah Laut in November 1979. This letter was a result of Mr M. Nuhuyanan having won the village head election that had been conducted several months before. Mr M. Nuhuyanan was now implying that his position was not the traditional village leader, but the modern village head.

Interestingly, Mr M. Nuhuyanan and his fellow *fam* members also developed an argument based on the tradition to account for his refusal to return his position. He argued that according to tradition, Yahaw Rahaded—the first traditional village leader—was not a patrilineal descendant of the Rahaded *fam* as explained by the story that Yahaw Rahaded was the illegitimate son of Balohoiwutun Balubun, who had 'illegally' impregnated Afenan Rahaded. Therefore, Yahaw should have been a member of the Balubun *fam*. In addition, Yahaw Rahaded was cared for by Bal Ulab Nuhuyanan who married Afenan's sister. Moreover, the Nuhuyanan *fam* believes that the appointment of Yahaw Rahaded was not based on 'real' adat. It was they argue, just because of his fluent Malay and his closeness to the Dutch.[13]

The conflict reached its climax in 1989. At that time, Mr A. Rahaded organised a meeting at the former village office which had been converted to a small prayer house because he and his allies refused to pray at the mosque. The group decided that Mr M. Nuhuyanan had to return the position before his retirement. A number of people were appointed to go to Mr M. Nuhuyanan's father's brother (who was the leader of Nuhuyanan *fam*) to discuss the proposal. Mr A. Nuhuyanan, the present village leader—who was at that time his father's secretary—was very upset and hit one of the representatives.[14] As a result, Mr M. Nuhuyanan's house, where Mr A. Nuhuyanan also lived, was attacked. Some people were injured and some parts of the house were damaged. An old woman who was the only person in the house died some days later. When the incident was brought to court, six of Mr A. Rahaded allies were sentenced to three months in jail.

Although the physical violence has stopped, the conflict continues. Mr A. Rahaded and his allies have written several letters to various government offices in the regency and Maluku province reporting Mr M. Nuhuyanan's misbehaviour. Although most of their letters were ignored, one of them brought Mr M. Nuhuyanan to court on charges related to the use of village subsidies. The regency court found Mr M. Nuhuyanan guilty of corruption. He was sentenced to six months in jail and fined Rp2.5m. However, the Maluku Province High Court in Ambon freed him when he appealed the case.

Conflict also occurred in the village head election which was held in 1992. There were two candidates—Mr A. Nuhuyanan (the son of the former village leader), and Ali Rahaded (Mr A. Rahaded's brother). Mr A. Rahaded's side almost won the election since the Kei Kecil subdistrict leader[15] supported him as had most of the *iri* in both settlements. The subdistrict leader supported Mr Ali. Rahaded because she believed that Rahaded was the *fam* who traditionally held the position. Most of the *iri* supported Mr A. Rahaded because he had promised to discontinue the tradition of rank at Dullah Laut, meaning that there would no longer be the noble and the former slave. Unfortunately, Mr A. Rahaded and his brother made a mistake. They issued a letter giving permission to a Madurese fisherman to dive for sea cucumber and signed it on behalf of the owner of Dullah Laut village territory and the village head. The Nuhuyanan

---

13 People believe that Yahaw Rahaded was the one who brought the Dutch to the Kei Islands.

14 Mr A. Nuhuyanan told me that his anger was also triggered by the fact that the person he hit was a slave who according to tradition, should not have been involved in the issue.

15 The Kei Kecil sub-regency leader is the direct superior of the modern village leader. In fact, it was the sub-regency leader who was in charge of the election of the modern village leader (see Chapter Two).

reported the matter to the subdistrict leader who considered the letter to be such an error in judgement that she switched her support to the Nuhuyanan. In her effort to support the Nuhuyanan, the subdistrict leader asked her staff to direct Christian voters to the Nuhuyanan's side. The subdistrict leader's staff member called the settlement leader of the Christian settlement and asked him to influence his people to vote for the Nuhuyanan. In return, the subdistrict leader's staff member promised the settlement leader that he would develop his settlement into an independent village. The settlement leader agreed. In the end, the Nuhuyanan won the election by fifteen votes.

After the election failure, the Rahaded faction did not give up. In fact, they did not acknowledge the Nuhuyanan leadership and based their rejection not only on tradition, but also on the law of village organisation. In a letter sent to various government offices, Mr A. Rahaded wrote that according to the law, traditional community is recognised. Therefore, he argued, the Nuhuyanan's leadership did not accord with the law.

In opposition to the Nuhuyanan, Mr A. Rahaded and his supporters ran their own programs which included building a small prayer house, constructing a stone dock, and widening their settlement (see Plate 6-1). These programs have been carried out on the eastern part of the Muslim settlement where most of the *iri* houses are constructed. Although several times after Friday prayer I heard leaders from Mr A. Rahaded's faction scolding those who did not participate in the communal work they organised, they felt satisfied with the development of their programs. Several new houses had been constructed and the stone dock was almost finished.

The history of leadership at the Christian settlement is as follows: from the leadership of the first traditional village leader until that of the fourth traditional village leader, the Christian settlement was led by a settlement leader from Yamko *fam*. In 1952, when the settlement was considered to be a village, his position was raised to that of village head. During the tenure of this village leader, the present settlement leader was his secretary. In 1984, the village leader passed away. His secretary—the present settlement leader—took the position until the village was merged with *desa* Dullah Laut Islam in 1989. This turned his *desa* into a 'child village' (*anak desa*) or a hamlet and demoted his position from a village head to a settlement leader.

Plate 6-1: Two young girls help with the construction of the stone dock.

Source: Author's photograph.

Conflict over the issue of leadership had never occurred at the Chrisitian settlement before which is quite interesting since the leadership of the settlement was not transferred 'from father to son' and the leadership position was never in the hands of a descendant of Yahaw Rahaded (the first traditional village leader). When I spoke with some elders, they told me that most of the first traditional village leader's descendants did not live in the village. They usually worked as government civil servants, teachers, and other white-collar workers outside of their village. Recently one family had returned to the settlement, but was not interested in the position of settlement leader because it was no better than the position they had retired from. Some other elders told me that Yamko, the first village leader, was one of the most important *fam* in Dullah Laut. In fact the centre of the village, which was located in their settlement, was named after their ancestor. Thus they believed it was appropriate to have one of them lead the village.

Concerning the present settlement leader, the elders argued that he was the one who knew best how a village or hamlet was organised because he had been involved in the business for decades. However, because the leader of Dullah Laut Village is Muslim, the people of the Christian settlement have developed a resistance movement. Although they have never expressed their resistance directly to the people or leader at the Muslim settlement, it was apparent that they had tried to cut their dependant connection with the Muslims. This movement was led by the present settlement leader.

Conclusion

Before I start my conclusion, I would like to summarise chronologically the main focus of this discussion, which was the raiding of the 'illegal' fishing company. The conflict started from an agreement made by Mr A. Rahaded with a grouper fishing company associated with a Taiwanese man and his Javanese wife. Their agreement was that Mr A. Rahaded would give the company permission to construct the company's base camp on Rumadan Island and operate their fishing activities in the surrounding waters. In return, the company would give a certain amount of money and employ some villagers for their operation. In this agreement, Mr A. Rahaded represented himself as the traditional leader and the lord of the land. The agreement was prepared with the involvement of the local army commander.

Some villagers raided the company when they were constructing their base camp at the location designated by Mr A. Rahaded. These people considered the operation of the company 'illegal' because its presence was without the permission of the legitimate village leader—the modern village leader. They felt that the company had not 'knocked at their door but had gone directly to a bedroom of their house'. It was agreed that the company would stop its activities and return to Tual until they had a further settlement with the village leader and a representative of the origin *fam*. The villagers took one of the company's speedboats as a guarantee that the company would comply with the agreement.

The company and Mr A. Rahaded, with the support of the local army commander, protested this incident the following night. This led to a customary meeting the following day which failed to reach any resolution and became an area of conflict between the three different political groups in the villages.

Now, the main aim of this discussion is to examine what the conflict was really about. This will be done by looking at the meaning of the conflict to each of the political groups in the village, particularly the leaders. I'll start by looking at the conflict from Mr A. Rahaded's perspective. It is clear that for Mr A. Rahaded, marine tenure represented the 'political capital' to oppose the power of the village leader. Leasing the right to use their sea territory to an outside fishing company was certainly a political ploy to demonstrate his leadership of the village. Regarding the agreement he made with the company, there were at least three crucial aspects to consider. First, leasing the village's territory demonstrated his territorial power both to the villagers and to outsiders. It was as if he'd said, 'Look! The territorial power over the village territory is in my hands, so I can represent the whole village in transferring the use right of the territory'. It was of course, a public challenge to the village leader's power on the issue of territorial rights.

The second aspect was the agreement with the company to employ villagers in their activities. This was considered effective for two reasons—reason one, the villagers became a buffer for both the company and Mr A. Rahaded. Whenever the opposing villagers confronted them, they were now able to say, 'look! This business is not only for us (the company and Mr A. Rahaded), but also for the villagers'. The implication was that the opposing villagers not only attacked the company and Mr A. Rahaded but also fellow villagers who, according to tradition, had use right over the village sea territory (see Chapter Five). As for reason two, the involvement of the villagers was also a means to win the hearts of the villagers. By Village Government law, the number of voters is important in securing a village leader position. Unless voted in by a majority of the villagers, a candidate for village leader is unable to take his seat. In the same way, villager(s) can remove a village leader from his position.

Money was the third important aspect of the agreement. Although few knew how much money the company was going to give Mr. A Rahaded when the incident took place, it was certain that money was a part of the agreement. This money was important for Mr A. Rahaded not only for his personal income, but for running his programs such as the settlement expansion, the stone dock, and the prayer house. In this context, the agreement was significant because the money involved was likely much greater than the amount generated voluntarily by villagers. There was also the possibility that the company would be asked for some additional economic support during the term of their agreement.

An additional value of great significance lay in the connections offered by the company. The most important of these was with the army. In Indonesia, the army was and still is a powerful institution. It not only controlled military related issues but also had significant influence in political, social, and economic affairs (Kristiadi 1999: 48; Crouch 1979). It was also an open secret that army officers not only used their power for the benefit of their organisation but also for their personal interests. This often led to their involvement in both legal and illegal businesses (Samego et al. 1998).

The circumstances in the Kei Islands were no different. Although civilian leaders led most of the government offices, the military's involvement in political and economic spheres was indisputable. For example, when I did my fieldwork, I met a military official who carried out a political census just months before the general election in 1996 which forced people to declared which political party they would chose. The census was considered a form of intimidation because those who did not chose the ruling party would be discriminated against. People also knew that some military officials were involved in some cyanide fishing businesses, yet no one—not even the head of the regency—dared to challenge their power (Adhuri 1998a).

Mr A. Rahaded was very much aware of this situation. He also knew how to use the military. Enlisting the support and involvement of the army in his agreement with the fishing company provided him with two significant advantages. First, the army became his shield from his fellow villagers' resistance. Those who opposed the agreement could be thought of as blocking the interest of the army. Second, his relationship with the army was also useful in smoothing his way whenever he met other local leaders. The latter was important since Mr A. Rahaded's final goal was to gain the position of modern village head.

From the point of view of the village leader's political faction, the raiding of the fishing company was a 'must' in order to demonstrate the village leader's power as the legitimate head of the village.[16] According to the village leader's supporters, the company's activities in Dullah Laut territory were 'illegal' for two reasons. First of all, this company had 'entered their house without knocking at the door,' as they put it. Without the consent of the village leader—the 'door' in the metaphor—no outsider could be allowed to make use of the village territory or resources for commercial purposes. People believed that both tradition and the Indonesian law regulated this. Second, judging by their equipment, the company was likely to use cyanide when fishing which was also illegal.

Mr A. Rahaded's agreement with the fishing company was also an 'illegal' way of challenging the village leader's power. According to his faction, the village leader was the only legitimate person who was authorised to represent the interest of the village to the outside world. They also believed that Mr A. Rahaded's claim of being the owner of the village territory was wrong. Dullah Laut territory was under the shared control of the origin *fam* which meant that Mr A. Rahaded had no right to make the agreement. In this context the incident was a way of 'getting things straight'. It would put the village leader into the top position and restore control over the territory to the hands of *Ohoiroa Fauur*. Ecological, economic, and social justice concerns raised in the customary meeting were only the means to justify the village leader's supporter's raid on the company. In fact, two years before the incident, the village leader allowed a different fishing company to use cyanide to fish in their sea territory for one year. At that time, it was Mr A. Rahaded who opposed the leasing.

---

16   As I mentioned earlier, the village leader was away when people raided the company. However, when I met the village leader in Jakarta and told him about this incident, he regretted that he was not in the village because he would have led the raid himself.

Finally, the most significant aspect of the incident in the eyes of the settlement leader was the timing. This incident occurred when the village leader was away and the settlement leader was appointed as the acting village head. From his perspective, he finally held the position that he had been dreaming of since Desa Dullah Laut Katolik had been transformed into the 'child village' of Dullah Laut and he felt that he could earn several points if he could succeed in persuading the customary meeting to allow the company to continue their activities. The first point was that villagers would perceive that he had passed the test of holding the power of village head, meaning that he would be considered a credible occupant of the position. Second, he could succeed in opposing the village leader without looking as though he was doing so by using Mr A. Rahaded, the army commander, and the company as a "smoke screen." Third, he could use the company's networks, as Mr A. Rahaded did, to assume power.

7. Marine Tenure and the Politics of Legality: Cyanide Fishing

> Government recognition and support of local resource management in coastal fisheries should be formalized.... In particular, explicit legal recognition needs to be given to the concepts of customary law (*hukum adat*) and local territorial rights (*hak ulayat*) (Bailey and Zerner 1992: 12).

> Because the basic customary marine law is still maintained and acknowledged by fishing societies, it is expected that basic customary marine law can be uplifted to be a provincial-level regulation which creates the legal and business certainty for fishermen so they can increase their welfare (translated from Lokollo 1994: 20–1).

The above quotations represent a popular recommendation for both central and local Indonesian governments to legalise the existence of customary marine tenure in Maluku and in Indonesia in general. In fact, this recommendation is one of the main elements in the creation of co-management (McCay and Jentoft 1996; Jentof et al. 1998) which refers to a management practice where government and fishing communities work together in crafting, implementing, and evaluating the policies related to marine tenure. Thus, such a recommendation is not unique to Maluku or Indonesia but worldwide.

There are two assumptions supporting this recommendation. First is that there is no government acknowledgement of traditional marine tenure, so far. And second, formal legalisation of traditional marine tenure will not only protect the practice from fading away, but most importantly create a better resource management practice.

This chapter will try to evaluate these recommendations by looking at the legal aspects of marine tenure in the Kei Islands and by placing formal and traditional legal provisions in the context of local practice. The former will be done by examining legal documents pertaining to the issue. A specific incident concerning cyanide fishing in Dullah Laut sea territory will be analysed to shed some light on the latter.

In this regard, contrary to the popular discourse on the subject, I would suggest that there is room to argue for the presence of legal recognition of traditional marine tenure by the government. As I will discuss in greater detail later, some articles in Indonesian law provide evidence that the Indonesian government recognises traditional marine tenure, and that support is even stronger at the provincial and district levels. In fact, referring to the 'illegal fishing' incident in the Dullah Laut sea territory, it was government agencies as well as police and military officers who forced the village head to address the problem with traditional rules and procedures. However, it was not the legality of the situation that led them to do so, but their interest in covering up the illegal use of cyanide for fishing which they were direct or indirectly involved. Cyanide fishing is against the fishery and environment laws of Indonesia.

Catching Cyanide Fishermen[1]

The Incident

Rumours of the presence of cyanide fishing on Dullah Laut Village's traditional fishing grounds had been in the air for about two weeks when the village leader and two villagers apprehended four cyanide fishermen on 2 August 1996. Cyanide fishing is the process by which fisherman squirt cyanide under stones or coral which temporarily stuns any fish in the immediate area. The stunned fish are then caught and stored in a holding tank after which fishermen depressurise them by puncturing their air bladders. Cyanide fishing is illegal under Indonesian government rules that prohibit pollution and the destruction of natural resources. Customary law also prohibits outsiders from commercial fishing in village fishing grounds.

The villagers were patrolling Dullah Laut waters when they spotted a foreign speedboat. When they approached, they saw a fisherman holding a hose in the water which was attached to an air compressor on the boat. This, they knew was a sign of cyanide fishing operations. The village head was very upset. He hit the offender and asked him to pull up the air hose. Another fisherman, wearing a wetsuit, was at the end of the hose. As the fisherman boarded the boat, the village head again lost his temper and slapped the fisherman. The village head asked the fishermen whether others were involved. They pointed to another nearby speedboat. An investigation revealed two more men, also using cyanide.

The four men and their boats were brought to the village. On board were diving gear and torches, as well as some pointed metal tubes about the size and diameter of drinking straws with wooden handles. The tubes are used to release the pressure from the distended air bladders of fish brought up from deep water quickly. There were also some live fish in a holding tank on one of the boats. One of the fish was a *Napoleon wrasse* which, by national regulation, is forbidden to be exploited for commercial purpose. Two cyanide pills were also found hidden on one of the boats.

---

1   This section has been published in Adhuri (1998a) and discussed in Adhuri (2001).

The fishermen said that a fishing company owned by a businessman in Makasar employed them. They also confessed to cyanide fishing. Clearly these four fishermen had violated the laws of both the Indonesian government and local custom. According to customary law, the apprehended men had stolen the fish, and by using a destructive fishing method, had also degraded the villagers' sea territory.

The village head decided to confiscate the boats and fishing gear. This is the customary action of a village head who, in this context, is considered the leading village official. Traditionally, a village head would only return confiscated items if certain customary procedures were followed. However, in cyanide fishing cases, one or more military officials often go to the village head and ask him to return confiscated goods. In these cases, the company gives a certain amount of money to the village head as 'smoke and betel nut' (traditional tokens of exchange). In this instance however, the village head tried to prevent this by making an official report to the government before any military intervention had occurred.

I accompanied the village head to report the case to the local police officer in the regency capital. We met the commander of the intelligence unit of the regency police post and the commander of the sub-regency police post. After we reported what had happened, the officers told us that this case was very difficult to prosecute. First, they said it was difficult to prove because they had no expert to examine whether cyanide fishing causes damage to the environment. We argued this point, but the discussion stopped when they told us what was the real reason for their reluctance to involve themselves in the issue, that 'we have a problem in prosecuting this case because our superiors are involved in this business'. They seemed to empathise with us but felt that they could do nothing. Nevertheless, they took the four fishermen to their office for questioning. They also suggested that we deal with the issue by means of customary law. This would put the village head in charge and prevent the involvement of government officials, including the military.

On 3 August 1996, an army officer from sub-regency army post came to Dullah Laut Village to 'invite' the village head to meet his commander.[2] The village head told me later that the fishing company had reported the case to the military post commander and asked him to persuade the village head to give the company back their speedboats and all of their equipment, and settle the problem 'peacefully'. In Indonesia, this almost always meant a request to drop the case. In return, the company will give some money to the village head. However, the village head refused his proposal. He argued that he had planned to report the case to the head of regency and it would be up to him to decide how to handle it.

---

2   The same military post commander was also involved in the conflict I discussed in Chapter Six.

Later that day, we went to the head of regency's house on Dullah Island. The head of regency responded to our report by saying that this case was not the first. He had known of such cases for some years, but it was a difficult problem. As an example, he told us about a case in a village where a military person was directly involved in the illegal activities. He told us under such circumstances, that there was nothing he could do because prosecuting the military about these activities was not within his authority.

The head of regency asked us to meet the commander of the regency army post. This seemed like an odd suggestion since we believed that our particular issue had nothing to do with the army. Later, the purpose became clear when I was shown a letter signed by the regency post commander (on behalf of the regency army cooperative's commander) and Mr A. Rahaded, the customary leader from Dullah Laut Village, who was discussed extensively in the previous chapter. The letter showed that Mr A. Rahaded had received an outboard engine from the regency post commander in return for the right to construct a base camp and fish cage and to operate a grouper fishing company in the village's territory.[3] The company was one of those that engaged in cyanide fishing and it also operated in official collaboration with the regency army cooperative. This made it seem likely that the head of regency had warned the regency post commander that the army's cyanide fishing operation was being challenged. When the regency head sent us to meet the commander was likely the way in which he informed him that his fishing operations were being challenged as well as gave the order for the commander to deal with the issue.

The head of the fishery office in the regency gave me a similar explanation when I questioned him concerning cyanide fishing. He told me that the involvement of Indonesian military officers had made the problem difficult to handle. However, it seemed as though he had found ways to benefit from this situation. One of his staff told me that he, in fact, was the local representative of the company that employed the fishermen we had caught. Moreover, the fisheries chief's brother told me that he personally had arranged all of the papers needed to export the catch.

I also found the company's licensing agreement to be unusual. The company had written a letter to the regency fisheries office asking for a letter of recommendation, which is one of the requirements that must be met before

---

3   Interestingly, there was no conflict in the village regarding the agreement between Mr A. Rahaded and the regency post commander. Some argue that this is because the fishing activities conducted by the regency post commander were the continuation of fishing activities by a company who had signed a contract with the village leader but went bankrupt before the contract finished. Others argue that it is because no one dares to challenge the regency post commander.

a fishing company is allowed to operate in the regency's water. In some cases, the letter of recommendation must be produced before a provincial or central fisheries office can grant a fishing license.

The letter from the company was dated 2 August 1996—the same day that the cyanide incident took place. The requested letter signed by the head of the Fisheries Office, was issued on 5 August 1996. So it appears that the operation had been unlicensed and that the letter had been requested so the company could use it if they were asked to produce a licence before a court. At the time he signed the letter, the head of the Fisheries Office could not have been unaware of the cyanide incident. He told me, in fact, that he had sent one of his staff to invite the Dullah Laut Village head to discuss it on 4 August.

When I told the acting commander of the regency navy post about the cyanide incident, he said that his post had only very limited resources. There were not enough speedboats and personnel to carry out patrols and it was therefore very difficult to observe illegal fishing practices.

The Customary Court

Frustrated by the lack of support from government officials, the village head decided to handle the case according to customary law. Customary law required him to arrange a customary court. At the time, village functionaries essential to the court were busy preparing local marriage ceremonies, and as a result he could not organise the court before all of the marriage ceremonies were over. But he was unable to ignore the fishing company which, through its representative and commander of the army post, was pressuring him to hold the customary court as soon as possible.

The customary court was finally held four weeks after the incident. An army official, a fishing company representative, and representatives of all origin kin groups in the village attended. After an opening speech from the army representative, the village head explained that the fishing company had violated both customary and government laws. 'For the latter, the case is in the hands of the officials in the capital', he said. 'Our concern in this meeting is the fact that they violated our customary law. It is our right to decide the fine for that violation.' Although the final court decision would be his, he said he would like to discuss the matter with all customary court committee members and ask their opinion.

A representative from the Christian settlement said that the company should pay ten million rupiah—an amount he said was prescribed in Indonesian law. This idea was not agreed on because it was not based on customary law. Mr A. Rahaded said that because the village head had beaten the fishermen, the company should decide how much they should pay. This suggestion was controversial because some court participants considered it to be benficial to the company. However, the suggestion was not unexpected because as previously noted, Mr A. Rahaded was the leader of the modern village leader's political opposition. He habitually denied the legitimacy of the leadership of the village head, led his own group of villagers, and ran his own village programs (see Chapter Six).

In addition, to support his position in the community, Mr A. Rahaded had tried to develop a good relationship with the military officials in the capital (see Chapter Six). He had also sought support from certain businessmen to run his programmes, for example, his agreement with the regency post commander. By criticising the village head by implying that his act of beating the fishermen was wrong, Mr A. Rahaded tried to show sympathy with the company.

Finally, the court agreed to fine the company six million rupiah and in return, the village head was to return all company property. This approach was due in part to the custom that villagers should not trouble outsiders in order to ensure their relatives who go or live outside their village will be treated well by others. In addition, the people needed money to continue construction of a church and mosque. Not wanting to be too hard on the company, the villagers decided that six million rupiah was a sufficient amount. This figure was not final, however. The company representative was asked to discuss it with his boss in Makasar. Another customary court would then be arranged to reach the final decision.

The court was held two weeks later, attended by the same people plus the army post commander himself. His presence was interesting because he ensured that, unlike the first customary court, the outcome of the second was carefully prearranged. Before the court was held, the military post commander, the village head, and the company representative discussed their plan. The company representative said that he had convinced his boss in Makasar to pay the six million rupiah.

However, the village head told me later that the army post commander had taken one million rupiah to be distributed among his friends. The village head was upset but he was powerless to refuse the army post commander. Interestingly, the village head also took two million rupiah and asked the representative of the company to say in the court that his company could only pay three million rupiah, the amount proposed by the company representative in response to Mr A. Rahaded in the first customary court.

When I asked the village head why he took the two million, he replied that this was not corruption but smoke and betel nut, which was his right as leader of the customary court. According to the customary law, he explained, it was the price of his effort in settling the problem. He also argued, 'Why should the army post commander—who had nothing to do with the case—be allowed to take one million if he was not also allowed to take a share?'

The court was run as planned. The representative of the company paid three million rupiah to the court. He also distributed Rp10 000 to each of the customary court committee as *uang alas meja* (table cloth money) as a token of appreciation for their attendance and contribution to the success of the meeting. Representatives of church and mosque construction committees were each given Rp1.5m. The case was closed when the village head returned the two speedboats and other equipment to the company representative.

The Legality of Communal Land Territory Rights

Unlike communal territorial rights on land, Indonesian legal scholars rarely, if ever, discuss communal marine tenure. Although some studies reveal that the practice of traditional control over sea territories was once widespread and is still practiced in some parts of Indonesia[4] it seems that Indonesian 'modern' legal thought is based on the 'European understanding that the seas are open to all' (Peterson and Rigsby 1998: 1). But following the discussions surrounding the legal position of communal or customary land ownership reveals that the legal issue of customary marine tenure is no less complex.[5] Although it is argued that communal marine tenure has never been formally acknowledged (see Panell 1997; Bailey and Zerner 1992; Warren and Elston 1994; Marut 2004), I will suggest that if we examine the laws and other legal documents closely, particularly at the provincial and district levels, traditional marine tenure is acknowledged to some degree.

The Indonesian constitution (art. 33, para. 3) states that "land, water,[6] atmosphere, and the natural resources therein shall be controlled by the state and shall be utilised for the greatest benefit of the people." This article, which is the basic legal reference for natural resource management in Indonesia, defines the state as holding the control of land and water (sea) territory. This article can also be interpreted to mean that the constitution does not acknowledge communal property rights.

---

4   See Pollunin (1984) for an historical account, and Adhuri (1993) and Wahyono et al. (2000) for contemporary practice.

5   See Haverfield (1999) to appreciate the complexity of legal acknowledgment of traditional land ownership and the need to incorporate the practice in the land reform package in Indonesia.

6   The term 'water' refers to inland water (such as lakes and rivers) and sea territory.

Nevertheless, the *Indonesian Agrarian Law 1960* stipulates that state authority to control land, water, and atmosphere can, in practice, be delegated to local government and customary law societies (art. 2, para. 4). It also states that the agrarian law applied to land, water, and atmosphere is *adat* law as long as it is not in contradiction with the national interest and the state (art. 5). These two articles while contradictory,[7] clearly demonstrate state acknowledgment of the community rights to land and sea. There is then a legal basis for the argument that customary marine tenure has been formally recognized by the state.

Now, I will examine the laws and regulations that specifically relate to marine territory and resources. If we examine fishery regulations during the Dutch period, we see that the customary rights of indigenous people were acknowledged. In 1916, the Dutch passed pearl shell and coral fishery regulations. Art. 2 of these regulations stated that:

> The right of indigenous people to fish [for resources] mentioned in article 1, is fully warranted; in all sea territories not more that five fathoms (nine metres) deep during low tide, indigenous people have exclusive right [to exploit the resources] if they have been making use of the territories since ancient time (translated from Anonymous n.d.).

Again, in *The Law of Coastal Fishery, 1927* (Kustvisserij Ordonnantie) the rights of local people were recognised. Article 6 of this law ruled that: 'those who do fishing according to this law will be allowed to do so only if they take into account the right of local people according to their *adat* and custom' (Anonnmous n.d.).

Unfortunately, the relevant provisions from the agrarian law and the Dutch fishery laws do not appear to have been used in developing contemporary Indonesian laws or regulations pertaining specifically to the sea. The *Fishery Law No. 9, 1985*, for example, has no article referring to communal marine tenure.[8] This law declares fishery management to be in the hands of the Indonesian government. This involves regulating all aspects of fishery operations including fishing gear, quotas, zones, licensing, and punishment for those who break the fishery rules.

---

7   The two articles are contradictory because one is based on the assumption that the traditional community law has no right over land, water, and atmosphere (article 2) while the other maintains that these three resources are governed by traditional law (article 5). According to traditional law, the communities own the land and waters.

8   The latest, *Fisheries Law No. 31 of 2004* and the *Law of Coastal and Small Island Management No. 27 of 2007*, explicitly aknowledge—and even respect and protect in the case of the latter—the right of traditional community. Article No. 6 (2) of the new Fisheries Law states that capture fisheries and aquaculture management should consider customary law as well as local wisdom when addressing issues that affect local community participation. The Law on Coastal and Small Island management stipulates that 'Government acknowledges, respects, and protects the right of traditional communities and local wisdom on coastal area and small islands which have been used for a long time' [article 61(1)]. Although these laws do not mention customary marine tenure, because it is a form of traditional law, one can argue that customary marine tenure has formal legal status. However, because the discussed incident took place before these two laws were passed, no reference will be made to these laws in relation to the discussion on the legal status of customary marine tenure.

It seems that the basic assumption of this law goes back to article 33 (para. 3) of the Indonesian constitution which states an area's resources shall be controlled by the state and utilised for the greatest benefit of the people, without opening the possibility of an interpretation that this right is transferable to customary law societies as stated in the *Agrarian Law of 1960*. Therefore, at this level we cannot find any legal basis to the claim that customary law societies have privileges in relation to their marine territory.

But this is not the end of the story. Looking at provincial fishery regulation in Maluku, one can argue that there are at least two indications that there is some recognition of customary marine tenure.[9] The first indication is that fishing companies—particularly those who wish to be involved in aquaculture or inshore fishery—are required to supply a territorial contract when they apply for a fishing licence. The second indication is that within the agreement signed by a fishing company when receiving their fishing licence, one article mentions that in operating their fishing activities, the company should respect the local traditions as they relate to territorial tenure.

Actually, the reason for incorporating a letter of territory contract was an attempt to address practical problems. In the 1970s, when pearl shell companies started their businesses in The Aru Islands,[10] there were many conflicts between the companies and local people. Being assured that they had a government licence in their hands, the fishing companies did not pay much attention to the local people, and driven by their belief that these companies were using their sea territory, local people protested their activities. This conflict was a serious burden for fishery offices in Ambon because they were in the middle of the conflicting parties. Although the fishing licences for these companies were issued by the central government, if there was a conflict at the fishing location, the central government officers did not know about it or more precisely, did not want to know. Therefore, the provincial fishery office was forced to handle any problems. Learning from this situation, the fishery office began to require a contract with the local people and respect for their traditions as one of the conditions upon granting a fishing licence to companies, a practice that resulted in customary marine tenure being officially acknowledged by the government's fisheries office.

We can find other evidence to support this argument if we check legal documents—particularly court decisions pertaining to conflict over sea

---

9 According to government regulation (*peraturan pemerintah*) *No. 15, 1990*, the governor or appointed provincial officer—in this case the head of the fishery office—may issue fishing licences for fishing companies located and operating in their administrative territory which use un-motorised vessels up to 30 gross tons (or 90 horse-power) and which do not involve foreign capital or workers.

10 The Aru Islands are an archipelago located on the eastern side of the Kei Islands (see Map 1-1). When I did my field work, this archipelago was part of Southeastern Maluku Regency. Now they form a different district.

territory—in both provincial and regency courts. The Southeastern Maluku Regency and Maluku Province high courts in Ambon have issued decisions concerning the boundaries between Sather and Tutrean villages and the distribution of the disputed sea territory (discussed in further detail in Chapter Nine). Both decisions clearly mention the ownership of particular social groups—in this case Sather and Tutrean villagers and the descendants of the original traditional village leader of Sather—over a particular sea territory which demonstrates legal acknowledgment of traditional marine tenure. It could be argued that these two decisions might not be considered to have legal status given the likelihood of an appeal to the Supreme Court in Jakarta. However, if we look at the case carefully, we see that the disputed issues were the boundaries and distribution of sea territory and not the existence of the communal right itself.

In qualifying the legal position of the above case, I came across another court document which detailed a conflict between the villages Wulur and Keli in sub-regency Pulau-pulau Kisar and Southeastern Maluku[11] over the Terbang Utara and Terbang Selatan islands and their surrounding territory.[12] The conflict was brought to the Regency Court of Southeastern Maluku in the early-1970s. In 1972, the regency Maluku Tenggara court issued a decision that was not accepted by the Keli villagers who appealed to the High Court in Ambon. In 1974, the high court ruled on the case but the ruling was once again rejected by the Keli villagers. They then brought the case to the Supreme Court in Jakarta but their appeal was rejected and the decision of the high court in Ambon was upheld (Indonesian Supreme Court, No. 1933K/Pdt/1992). This decision was executed in 1986. Part of the decisions reads as follows:

> To conclude that Terbang Utara Island and Terbang Selatan Island with their *meti* [sea territory] located in southern part of Damer Island, Pulau-pulau Kisar sub-regency are the property of all Wulur villagers, [these two islands and their *meti*] are the territory of *petuanan* Wulur village. (emphasis added, translated from the decision of Maluku High Court, 16 July 1974 No. 113/1973/PT/Perdt)

This decision, which makes specific mention of *meti* and *petuanan*, confirms that in legal practice, communal marine tenure is acknowledged by the Indonesian legal system even if formal acknowledgement of traditional marine tenure is lacking in Indonesian laws. The result is a considerable degree of legal ambiguity on the issue of customary marine tenure as there also is in relation to communal tenure on land.[13] Some may argue that this ambiguity makes it hard

---

11 Like the Aru Islands, these islands were part of the Southeastern Maluku when I carried out my research. Since 1999, they have become part of a new district called Maluku Tenggara Barat.
12 See Panell (1993) for an account of this conflict.
13 See Haverfield (1999) regarding communal land tenure.

for communities to depend on their right to the sea, but I would argue that the ambiguity opens the way for more options, allowing local parties to use either laws or communal marine tenure to adequately address an issue.

In the previous section, I describe how local police and government officials forced the village head to drop the case and handle the issue by means of customary law. The village head did so and an agreement was reached. This course of action meant that the customary law, particularly the traditional law of marine tenure, gained practical legitimacy. Drawing on my discussion of the practical legal use of customary marine tenure, we can say that the decision of the customary court has formal legal ground as well.

However, was it legal reasoning that caused customary marine tenure to gain legitimacy in this case? I am afraid I would answer 'no' to this question. Neither the community nor the local military officers and bureaucrats used legal logic when choosing which avenue to follow in pursuing the case. When the local bureacrats forced the village head to drop the case and choose the customary law option, their motivations were primarily pragmatic in that they wanted to avoid conflict with the district military commander who was involved in the same business. In conclusion, I would argue that the formal legal definition of customary marine tenure is not the most important one. In the case of the cyanide fishermen operating in Dullah Laut's waters, people did not care about the formal legal definition of customary marine tenure. Its legal recognition was the unintended side-effect of a pragmatic approach to local resource management.

The Illegality of Cyanide Fishing

Looking back to the customary court, we might notice that the court only discussed the fact that the fishermen had violated the traditional law of the sea. In his opening remarks, the village head told the audience that the fishermen had used cyanide and that this had been reported to government officials. Yet, there was never any talk about this issue between the Indonesian officials and no formal action was taken against the company. Does it mean that cyanide fishing is a legal practice?

There are many laws and regulations that define cyanide fishing as illegal which means that the fishermen and companies using cyanide should be subject to punishment. For example, the Fishery Law stipulates:

> All persons or companies are prohibited to fish or do aquaculture using material and/or tools that endanger the sustainability of the fish resource and its environment. (translated from *Fishery Law No. 9, 1985* art. 6, para. 1). Anything that might cause pollution and damage to the

fish resource and/or its environment is forbidden. Violation of one or both articles is subject to ten years maximum in jail and/or a fine of one hundred million rupiah (ibid.: art. 7, para. 1).[14]

Now, one might ask whether cyanide fishing causes pollution or damages the environment, as the police did when the village head and I reported the case. Indonesian waters were not the first to fall victim to cyanide fishing—the Philippines suffered from the practice as early as 1960. In the Philippines, there were reports that cyanide use had not only stunned the targeted fish but killed smaller fish, fish fry, invertebrates, and coral reefs and even caused skin diseases for people who were exposed to it on a regular basis (Rubec 1986, 1988; Dayton 1995; Milan 1993).[15] This sort of information as well as widespread local knowledge about the negative impacts of cyanide should have been enough to support the claim that cyanide fishing was damaging to the environment, which should have motivated the police and other legal officers involved in the case to search for stronger proof of local cyanide fishing operations.

In addition to the cyanide fishing violation, the company could also be arrested on at least two other grounds. First, they caught *Napoleon wrasse* which was banned by the Minister of Agriculture in its decision No. 375/Kpts/IK.250/5/1995. Second, the company had operated without a fishing license. As I mentioned in the section describing the incident, the company was in the process of requesting a letter of recommendation from the local fisheries office. This means that they did not have the fishing license that was required by the law (art. 10). This violation might have sent those who were responsible to prison for between two-and-a-half and five years, or led to a fine of between twenty-five to fifty million rupiah.[16]

To contextualise this incident, I will now refer to the formal discourse on conservation and economic sustainability. This national discourse should have led us to expect that local police and bureaucrats would bring the fishermen and the company to court. Since the late 1970s, the issues of environmental conservation and sustainable development have gained popularity in Indonesia due to demands from external and internal agencies that the Indonesian government pay greater attention to these issues (Warren and Elston 1994: 7; Zerner 1994b: 1100; Hardjono 1991). To external donor agencies, attention to environmental issues and sustainable development became an increasingly important condition for the receipt of aid funds. Internally, the emergence of a middle class with a growing awareness of environmental issues—demonstrated

---

14 A new fishery law (No. 31, 2004) states that the same violation is subject to six to ten years maximum punishment in jail and a fine of Rp1.2–2 billion [article 84 (1–3)].

15 See also Johannes and Riepen (1995) for the Asia-Pacific region.

16 According to the new fishery law (No. 23, 2004), this violation carries eight years maximum imprisonment and a Rp1.5 billion fine (ibid.: art. 93).

by the flourishing of NGOs—put more pressure on the Indonesian government. In addition, practical issues such as conflicts over land tenure and problems associated with environmental degradation forced the Indonesian government to develop laws and regulations which were meant to prevent people from abusing the environment.

Nevertheless, in the case study I have discussed, conservationist discourse and the formal 'illegality' of cyanide fishing were not strong enough to drive local police to deal with cyanide fishing as an illegal activity. The case demonstrates that the formal legal definition of a particular issue will not be automatically applied in real situations.

Conclusion: The Politics of Legality

The discussion on the legality of traditional marine tenure and the illegality of cyanide fishing shows that the 'practical legality' of the former emerged out of the ambiguity of its formal legal definition, while the 'practical legality' of the latter directly contradicted its formal legal definition. What both examples show is that the formal legal definition did not really count. What mattered were interests and power.

It was in the best interests of all parties—locals, bureaucrats, police etc.—that traditional marine tenure and its associated customary procedures be considered legal. Bringing the case to court within the formal legal system would have threatened their interests. The police and the regency head were keen to maintain good relations with the commander of the military district post (*dandim*) and other military officers from whom they may require support in the future (as they had in the past). The district commander and the head of the fisheries office had a direct economic interest that would have been threatened by a formal court case. For the villagers, it was in their best interest to deal with the problem according to customary marine law. In fact, the villagers received a twofold benefit from holding the customary court. The first benefit was the money paid by the fishing company, money they would not have received without the customary court. The second benefit arose when the police suggested that they resolve the issue by means of a customary court and the military and local bureaucrats allowed them to do this. This course of action might be considered a good precedent for future cases.

In terms of the power structure, as shown in Figure 7-1, the district military post commander was at the top. Although formally he was on the same level as the regency head, as the top-ranking military leader in the regency, his power was uncontested.[17] It was obvious that the regency head and the police officers were forced to ignore cyanide fishing despite their objections to it. Since they had no choice, they—however indirectly—pressured the village head to accept the circumstances, and because the village head was at the bottom of the structure, he had no power to refuse what he was told to do.

District level

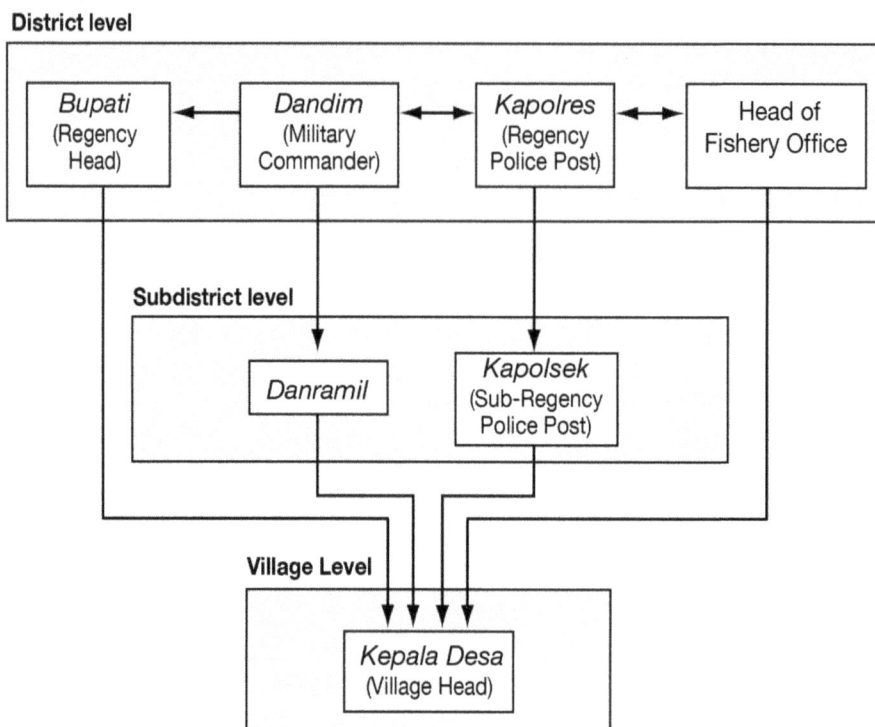

Figure 7-1: The power relations of the parties involved in the cyanide fishing incident.

Source: Author's fieldwork.

Since power and interests define legality, we might then question the effectiveness of formalising traditional marine tenure in the Indonesian legal system. In fact, regarding the unbalanced distribution of power, the formal recognition of marine tenure might only add to those who hold power. The result, as indicated by the case discussed, may disadvantage the powerless and further environmental degradation. Therefore, sustainable and socially just marine resource management might not be as clearly visible in the future as advocates of formal recognition of customary tenure suggest. In saying that,

---

17   During the New Order regime, military forces and policies were extremely powerful. They controlled Indonesia socially, politically and economically (see for examples Vatikonis 1998: 60–91; Samego et al. 1998; Crouch 1979). The case that I am discussing is an example of a general trend throughout Indonesia.

however, I do not mean that the formal acknowledgment of the traditional marine tenure is not important. In various situations, increasing the bargaining position of local communities through legal recognition will play an important role, but we should also be aware that there are some requirements needed to ensure that the legal construction is effective. One of these requirements is relatively equal power distribution between all parties associated with the resource. Unfortunately, we do not find such equality in the community we are discussing.

8. The Economy of Marine Tenure: The Clove Season Incident

The introduction and integration of the market economy into local communities is often identified as one of the causes of the breakdown of traditional sea management practise. For example, Johannes (1978, 1981) observed that the cash economy introduced by Westerners had degraded traditional marine tenure in Oceania. He argued that the introduction of a market economy encouraged competition for cash which led to increased exploitation of marine resources through the adoption of more effective and efficient fishing technologies and techniques. 'Under such conditions', Johannes (1978: 357) wrote, 'a conservation ethic cannot thrive. Conservation customs practiced voluntarily by the individual erode first'. He notes that 'the most widespread and most important single marine conservation measure employed in Oceania, and the most important, was reef and lagoon tenure' (Johannes 1978: 350).

Hviding proposes an alternative argument. He suggests that:

> Even where a fairly open access to fishery resources seems to be the rule, the commercialisation and intensification of marine exploitation may lead to a sudden upsurge of a multitude of ideas about customary boundaries and fishing rights—seemingly from nowhere. Hviding (1989: 5–6)

This argument looks at marine tenure as a response to the commercialisation of fishing activities that is usually associated with the adoption of a more advanced and efficient technology. We might say that this is another version of Polunin's (1984) theory, which suggests that the absence of marine tenure might be attributed to the economic insignificance of the resources, creating no incentive for investment. When the market 'informs' people of the value of the resources, boundaries are drawn and claims are laid down so people can exclude others from gaining benefits from their territory.

Discussions of marine tenure in Maluku provide some support for the first theory. For example, Bailey and Zerner (1992) and Nikijuluw (1994), argue that the commercial economy, represented by commercial marketing, fishing companies, and the adoption of advanced technologies, has marginalised the practice of traditional marine resource management in Maluku. They suggest that trade and government promotion of exports have forced villagers to shorten the traditional closed season, allowing people to fish and harvest other resources more often. They also argue that private entrepreneurs have taken control of people's resources and territory through rental arrangements or by entering and extracting resources from people's territory. The villagers were observed to accept this situation because they became economically dependent and politically powerless.

This chapter will discuss the relationship between commercialisation and marine tenure in the Kei Islands. It will focus on the influence of international trade in frozen anchovy. The discussion will develop by analysing conflicts in Dullah Laut over sea territory. In particular, I will discuss a conflict between Dullah Laut villagers and Ut and Selayar villagers. The conflict was triggered by the operation of a lift net[1] by Ut and Selayar villagers in Dullah Laut territory.

The conclusion I have reached on marine tenure relates more to that of Hviding than to others who have written about marine tenure in Maluku. While I disagree that marine tenure emerges 'seemingly from nowhere', I argue that the international market in fishery resources has prompted people to strengthen their existing traditional marine tenure practices. Traditional boundaries have been emphasised and the principles of communal marine tenure have been reinforced. Since traditional marine tenure is a contested practice, the international market and the strengthening of customary marine tenure have created conflicts within and between community members as well as outsiders fishing in the territory.

The 'Clove Season' Incident

The 'clove harvesting season' (*musim cengkih*) in Dullah Laut is unique. It confused me the first time an informant told me about it. In its common understanding, the term does not only refer to the presence of clove trees but also an abundance of cloves. People would not use the term season (*musim*) in relation to something that was only small in number. Yet, I had never seen even a single clove tree during months of fieldwork. When my informant told me that the cloves in Dullah Laut did not grow on the land but swam in the sea, I started thinking that this must be a metaphor. Finally, he explained, 'We are not talking about the clove spice; this clove is trevally (*Carangoides* spp, *bobara*)' (see Plate 8-1). He explained that every year, thousands of trevally—marine fish—make their way into Dullah Laut sea territory. When there are so many in number, they are easily caught and money is easy to earn. Even for those who fish for fun, this is a particularly joyful time (see Plate 8-3). For most fishermen, the fun

---

1 A lift net is a floating rectangular structure equipped with a net that can be sunk into the water and then lifted using a rolling device. Its four corners are anchored in order to immobilise the structure. When the fisherman wants to move the lift net to another fishing spot, he will take the anchors out of the water and use a boat to tow the lift net. When operating, lift nets use lights (carosene lamp or mini electric generator) to attract a school of fish which are then trapped in the net once the net is lifted (see Plate 8-2).

of fishing for trevally is the best fishing experience they have all year. Therefore, he concluded, the trevally fishing season was a time of high excitement, like the excitement of the Ambonese and people in other parts of Maluku when they harvest the clove spice.

Plate 8-1: The harvest of musim cengkih or clove harvesting for trevally, the Bobara (*Carangoides* spp).

Source: Author's photograph.

I had been waiting to experience this excitement when finally the 'cloves' were ripe. I was at the fish market in Tual when a villager told me that the trevally had 'played' in Dullah Laut sea territory. That day, I bought new fishing equipment and prepared a boat with my friend and the Dullah Laut Village head. We also asked another friend to join us so that we could share the cost of the fuel. The village head and I went back to the village to get more detailed information on the location of the 'clove season'. That evening, the village head together with some other villagers, told me about their past experiences of trevally fishing. The trevally's large size and the shallow water where they congregate make it very hard to land the fish. They jokingly expressed their doubts that I would be able to successfully pull the fish up into the boat.

Plate 8-2: The Bagan (lift net).

Source: Author's photograph.

Plate 8-3: The joy of *musim cengkih*, fishing for trevally.

Source: Author's photograph.

Finally, I got the chance to be involved in the fishermen's joy of the 'clove season.' The village head and some of my friends collected me at about five in the morning the next day. It took us about 15 minutes to reach the fishing site at Wada Iyuwahan. The sea was very calm when we arrived at the location. I saw about 50 boats of various sizes on the fishing ground. People—one or two in small boats and up to five in bigger boats—were ready with their fishing lines. Everyone was talking and shared jokes while waiting for the fish to take the bait—'meal time' according to local expression. As the sun rose, thousands of trevally came into view at approximately three to five fathoms depth. Although no one had yet caught any fish, I already felt the excitement and the others shared that same feeling.

At around six in the morning, the trevally began to take the bait. People screamed excitedly as the fish pulled on their lines. Within minutes, I observed that almost all of the fishermen had a turn pulling a fish out of the water. Looking at how they pulled on their lines, I believed what the village head had said about the fishing being a real challenge. The fishermen had to pull hard on their lines and many were forced to loosen their lines as the fish ran in the opposite direction. There was considerable chaos as lines became tangled in fishermen's hands or with other fishing lines. When I saw the fish hauled aboard the boats, I had to acknowledge that the village head had not exaggerated about their size.

This was a big day for everyone. By about eight in the morning, a middleman from Dullah Laut had already collected approximately 300 fish, worth almost a million rupiah. Around midday, he had collected a similar number again and then took them to the fish market in Tual. At two in the afternoon, he made a third trip selling the fish to the market. Thus, the fishermen shared a total catch worth around three million rupiah in that single day. The middleman got approximately Rp500 000–750 000 profit from his sale of the fish in Tual. The village leader, myself, and two friends in our boat caught 40 fish which netted us Rp120 000, enough to cover the cost of fuel, lunch, and some boxes of cigarettes for that day and the day after. We went back home around four in the afternoon with an agreement to fish again the following day.

The second day of the 'clove season' was disturbed by the presence of a lift net. People complained that this lift net would catch all the anchovy which people believed was the reason for the trevally's presence. This concern was reasonable since people had seen the lift net catch tons of anchovy in one day based on the activities of a speedboat that transported the anchovy from the lift net to the fishing company's ship. On that day, their speedboat made five trips, each time with a load of approximately 700 kilograms.[2] The village head took direct action. He sent his uncle to ask the lift net fishermen to move their lift net from the current location.

On the third day, the conflict escalated. There were three additional lift nets in the area. Although still abundant, the fish did not take the bait causing the fishermen's catch to decrease significantly. Expressing their anger, some of them threatened to cut the lift net anchors. Some others accused the village head of taking bribes from the lift net fishermen because he did not do anything to get rid of them. This claim was addressed by the village head's uncle who had warned the lift net workers the day before. When the village head arrived, he also proved that he did not support the lift net owner. He ordered people to cut the anchors of all of the lift nets. However, as he finished speaking, a fisherman from Ut Island replied by saying, 'Those who cut the anchor, his neck will be cut too'. A fisherman who was a descendant of the war commander replied, 'Let's go to land and see who cuts whose neck'.

Apart from the fact that the lift net operations disturbed the clove season, Dullah Laut villagers considered the lift net fishermen's presence illegal because the lift nets were cast without the consent of the village head. This meant that the lift net owners had violated customary marine tenure which requires that those who wish to fish for commercial purpose should seek permission from the owner of the territory. They also suspected that the people of Ut Island were claiming ownership over the location where the lifts net were operating. The threat made by a fisherman from Ut Island marked his association with the lift net owners. From this, people suspected that the Ut villagers had granted the lift net owners the right to fish in the fishing spot and based on the rule of customary marine tenure, this meant that the people of Ut Island were claiming ownership of the territory. This suspicion was proven true when Mr H (the brother of the Ut fisherman) met the Dullah Laut Village head who was fishing with the son of the former Dullah Laut Roman Catholic village head. He told the village head that it was the leaders of Ut Village who had allowed the lift net fishermen to set their lift nets in that location. In addition, he mentioned that the territory where the lift nets operated—called Metan Er—was their territory. In support for his claim, Mr H chanted a customary song mentioning his ancestors' involvement in the Papuan War (see Chapter Four). Mr H said that the 'three houses' (Rahan Itel) consisting of Yamko, Henan, and Rahaded *fam* of Dullah Laut village, had given Metan Er to his ancestors as a sign of gratitude for their help in the war.

Responding to the claim, the Dullah Laut Village head said that he was not opposed to the operation of the lift net fishermen in that area because he

---

2 This estimate was based on the number of plastic boxes containing anchovy. Each box contained approximately 100 kilograms and the speedboat transported seven boxes on each trip.

understood that the money paid by the fishermen would be used for the construction of a mosque on Ut Island. However, the village head said that the Ut Island people should have informed him, since the lift nets operated in Dullah Laut village territory. This was the basis of his objection to the claim. He also stated that the name of the location was not Metan Er but Wada Iyuwahan. The son of the former Dullah Laut Roman Catholic village head supported this: 'It is true that people of Ut Island have helped our ancestors in the Papuan War but they were only given use right over mangrove trees on Baer and Ohoimas islands'. Mr H took exception to this point. He insisted that his version of the 'history' was correct version. The Dullah Laut village head finally told Mr H to come to Dullah Laut Village and solve the matter before a customary court of the origin *fam* (*Ohoiroa Fauur*). Mr H left our boat without saying a word in response.

The following day the conflict became even more serious. Contrary to the villager's request, the lift net fishermen kept fishing and brought in even more lift nets. On the fourth day, five additional lift nets were towed to the location and the Dullah Laut village head received a letter signed by Mr Dullah Rumagiar (Mr H's brother) on behalf of the Ut Island leaders which raised two points: first, was that the people of Ut had allowed the lift net fishermen to operate in Metan Er, the location of the trevally; and second, they re-stated their claim to possess Metan Er based on the narrative Mr H had uttered to the village head. For Dullah Laut villagers this was a challenge of war because they believed that defending their territory and family members were the most important obligations for which they would sacrifice their wealth, slaves, or even lives if necessary.

In response to this matter, the village head held two customary meetings, one at the Christian settlement and another at the Muslim settlement. Except for Mr A. Rahaded (see Chapter Six), all representatives of origin *fam* in both settlements participated in the meetings. Both meetings discussed the nature of the problem and strategies to deal with it including evaluating the narrative of origin that explained the relationship between the people from Dullah Laut and from Ut Island. In this regard, all agreed that there was only one version of the origin narrative—the one that was told by the son of the former Dullah Laut Roman Catholic village head. They also concluded that the location of the 'clove season' was Wada Iyuwahan and not Metan Er. They decided to meet the lift net fishermen out at sea on their lift nets and ask them to leave. The lift net fishermen would be given one day to comply with the demand. It was also decided that the village head would prepare a letter rejecting the claim of the Ut Island people. The letter would be directed to the top leader of Ut Island, the Kings Baldu[3] and Tufle,[4] and government officials at the subdistrict level.

On the day agreed to raid the lift net fishermen, some guests visited the village head early that morning. The first guest was the owner of a fishing company that bought anchovy from the lift net fishermen, and her brother and nephew. They confessed that their ship had towed five of the lift nets but that their company did not own them. They said that their ship had towed the lift nets only because their company bought the anchovy they caught. According to these guests, they had nothing to do with the operation of the lift nets in Dullah Laut territory. However, they asked the village head not to harm the lift net fishermen.

The second guest was Mr O, a Dullah Laut villager with a Butonese father. The aim of his visit was to convey a message from his fellow Butonese, the father of the first lift net owner. According to Mr O, the Butonese fellow had apologised for what had happened and had planned to meet with the village head to ask permission to set his lift net in Dullah Laut sea territory, but his visit was postponed because his brother passed away. Meanwhile, his son had brought his lift net to Ut Island in order to get closer to Dullah Laut where he wanted to fish and his son's wife's father, one the leaders of Ut Island, told his son that it was all right to set the lift net in place. In fact, it was his son's father-in- law who had towed the lift net in place. Mr O also added that the owners of the other lift nets were fishermen from the same village as the first lift net owner and were financed by an anchovy exporter.

The visits of these guests brought about a new understanding of what had happened. It became clear that the operation of lift netting in the 'clove season' location was the idea of an Ut Island villager, not all of the leaders in the village or the head of the village. Initially, the 'permission' applied only to his son-in-law. The other lift net fishermen just followed their fellow villager to operate in the same location. It was not clear whether they had spoken to the father-in-law of the first lift net owner or not. However, it was clear that the fishing company was directly involved in this incident as admitted by the fishing company owner, who said their ship had towed the lift nets.

This new understanding did not alter people's plan to raid the lift nets that day. At eleven in the morning, just after the Christian villagers had finished their Sunday prayer, around 20 villagers led by some origin *fam* leaders went to the lift net's location. They met the fishermen and told them they had one day to take their lift nets away. If they had not observed this request by the next

---

3   Baldu is the king of Utan Til Warat domain to which Dullah Laut is attached.
4   Tufle is the king of Tual domain to which Ut Island is attached.

day, they were told that no one would be responsible if something happened to them. This was clearly a threat that people would harm the fishermen or their lift nets if they did not comply.

The raiders had predicted that they would meet some resistance from the fishermen and brought knives and other weapons with them for protection. Even though there were now 11 lift nets operating and the owners and workers out numbered the raiders, they did not argue with or confront the raiders physically. They accepted the request and promised to move their lift nets out of the territory that day.

The Economy of Village-Based Fishing[5]

In examining the 'clove season' raiding incident and other conflicts pertaining to the Dullah Laut sea territories, my belief is that fishing competition between villagers and outsiders is one of the main issues that has triggered conflict. This is a complex issue which involves not only who can fish in a particular sea territory, but also the level of acceptable exploitation which is closely related to fishing technology and access to marketing. The following will outline the economics of fishing activities conducted by Dullah Laut villagers and outsiders.

Dullah Laut villagers mainly used four types of fishing technologies: fish pot, stake trap, line, and gillnet.[6] (See Plate 8-4) The fish pot, stake trap, and line methods of fishing are considered the 'real' traditional fishing technologies still in use. Their use may be as old as the community itself or at least several generations. Gillnets is a relatively new method and was first used in the late-1970s coinciding with the introduction of outboard engines.

Although some villagers—such as net owners and the elderly—used only a particular fishing technology, most fishermen in Dullah Laut used two or more types of fishing gear. It was not unusual for villagers to retrieve their fish pots in the morning and go line fishing at night. Some villagers even used three fishing technologies in the space of 24 hours. In the morning they would dive to collect their catch in the stake traps and fish pots and send them to the fish market in Tual, or sell them to a middleman in the village. During the daytime, they would prepare their fish pots to set out in the afternoon before line fishing at night.

---

5   A detailed account on the economy of fishing is presented in Appendix 1.
6   Bubu (*fuf*) is a box-like fish trap made from bamboo. Stake trap (*fean*) is a fence-like fish trap, also made mainly from bamboo. The lift net is a rectangular floating device equipped with a net that is lifted when the target fish have schooled on top of it. Pressure lamps onboard the boat are used to attract fish to gather above the net.

Plate 8-4: Two fisherman prepare the bubu, or fish traps.

Source: Author's photograph.

Investment and Income

In terms of costs, traditional fishing technologies only require a relatively small investment. The net technology has a considerably higher initial capital cost for the net and boat because the boat is not only used for fishing activities but for other purposes as well, such as transportation. Typically, a fisherman could go to sea with only Rp5000 for buying some line or other fishing requirements. For a fish pot fisherman, about Rp60 000 buys 15 fish pots, the minimal number for a proper operation. For stake trap fishing, a fisherman would need about Rp250 000 to have a stake trap ready for operation. In comparison, a gillnet fisherman told me that he bought an outboard engine for Rp2.6m in 1986 that he used for net fishing. Needless to say, several hundred thousand more were required for the net and a bigger boat to service the new outboard engine.

I observed at least five ways fishermen obtained capital for their fishing activities. The first was from operating their own fishing business or selling agricultural products. The second was from borrowing the money from a village middleman to whom they sold their catch. They usually agreed that the loan repayments would be deducted from the price of their catch. This kind of arrangement was not strict in terms of the level of repayments or the period of the debt. The third way was to form a group in which one of the members provided the money to buy the fishing gear while others contributed their labour to operate the business. The fourth was to obtain a gift or a loan in the form of money or fishing gear from non-business connections such as relatives or friends. The fifth way occurred by buying fishing gear directly from a fishing gear store in the capital city of Southeastern Maluku District and establishing a mortgage system of repayment. Most of the money to pay for the gear would have come from sources within the community on Dullah Laut and only this scheme would involve non-villagers contributing to the investment.

In line with the investment, the overall income generated using the traditional fishing gear or gillnets was relatively low. Data collected on fishing catches revealed the following monthly net incomes of nine fishermen (see Table. 8-1).

Table 8-1: Monthly net income for Dullah Laut fishermen, 1996.

Technology	Net monthly returns (Rp)	
Fish pot (fuf)	204 166	193 000
Stake trap (fean)	264 133	
Line	267 583	283 833
Net (crew)	171 209	139 018
Net (owner)	659 178	330 564

Source: Calculated based on data presented in Appendix 1.

As shown in Table 8-1, the net monthly incomes ranged from Rp171 209 to Rp659 178. Except for the net owners, the income using any of the traditional fishing methods was similar. That said, the traditional technology providing an average income of around Rp230 000 is more lucrative than crewing for the gillnet owner.

Fish Marketing

All of the fish caught using the techniques described were sold locally. The first marketing method was to sell the catch directly to consumers in Dullah Laut, in other villages, or at a fish market. The second technique was to sell the catch to a middleman in the village or in the Tual fish market. A third was to sell the catch to a middleman in Dullah Laut Village who then sold them to another middleman in the fish market who sold them to consumers in the market. The second and third ways generated employment for six people in Dullah Laut and around 20 people at the fish market in Tual. The first not only sold the catch of Dullah Laut fishermen, but also the catch of fishermen from other villages. Figure 8-1 describes these marketing schemes.

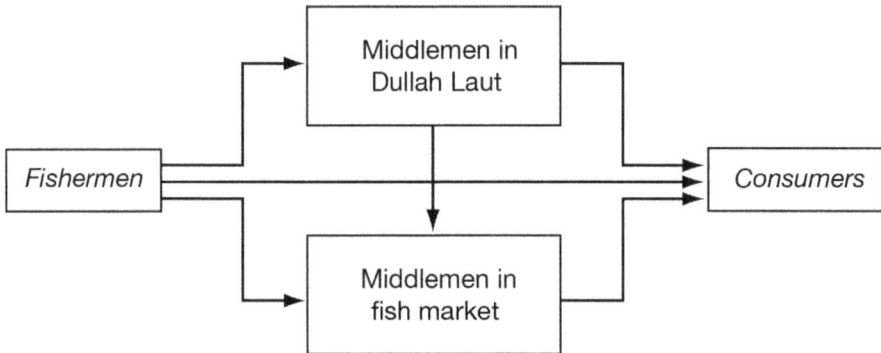

Figure 8-1: The domestic marketing network.

Source: Fieldwork research.

The fish marketing business provided a living wage for about six to eight people in Dullah Laut Village. Most of them were middlemen in the Muslim settlement, and only two were from the Christian settlement. I recorded the net monthly income of four of these middlemen and their earnings to be: 172 383; 160 633; 245 750; and 936 569 rupiah per month. It seemed the significant differences in income between the first three and the fourth were caused by their way of acquiring and selling the fish. The first three fishermen ran their business in the same way—they bought the fish in Dullah Laut Village, and sent the fish by public sea transportation to the fish markets where they were sold. So, the quality and quantity of the fish they bought and sold was determined by the fishermen coming to the village from their fishing trips, and the departure time of the public sea transportation. The last middleman operated his business using a boat with an outboard engine. This technology made it possible to collect the fish from the fishermen while they were still out at sea. It also made it easy for him to send the fish to market any time he had enough fish to sell.

The Economy of Export Oriented Fishery

The discussion of export oriented fishery in Dullah Laut sea territory should cover both live reef fish and anchovy fisheries. However, since the former is not related to the clove season incident, I will only discuss anchovy fishing.

The anchovy export in the Kei Islands is associated with the use of lift net technology, which is the only technology suitable for catching anchovy. Lift net fishing has been popular since the 1980s and was created in 1964 when a fisherman in Sathean Village on Kei Kecil Island modified his mosquito net as a lift net. The net only lasted a couple of days and his fellow villagers protested about its use at the time (Adhuri 1993). It was not until 1983 that a second lift net was constructed and operated. Although it generated conflict amongst Sathean villagers, they managed to resolve the conflict and continue using the lift net. When I did my field research, lift nets also operated in some other villages such as Selayar and Ngilngof on Kei Kecil Island.

It was only in 1995 that lift net fishing in the Kei Islands became connected to the international anchovy market when a Taiwanese businessman bought anchovy from local fishermen for export to Taiwan. His business started in Ambon in 1993 when, in collaboration with a fishery company in Jakarta, he brought two ships to Ambon and started buying and exporting anchovy. In 1995, he sent one of his ships to the Kei Islands because the supply from lift net fishing in Ambon was well below his export target.

When he began his business in the Kei Islands, he did not involve himself in fishing activities at all. He depended on supplies from the village-based lift net fishermen. However, when he realised that the resources were quite promising—and production still below his export demand target—he encouraged more fishermen to become involved in lift net fishing. In 1996, he adopted a loan scheme in which he loaned money to those who were interested in lift net fishing. In return, those who borrowed the money would sell the catch—particularly anchovy—to the company. Special arrangements for the selling price and the way in which the fishermen would pay back the company would be agreed upon by the company and the lender. There were some local businessmen who wanted to get into the anchovy business and some of them signed up to the loan scheme developed by the Taiwanese businessman as well. Other local businessmen constructed their own lift nets but asked local fishermen to operate the nets for them.

The capital needed in lift net fishing was higher than for all other technologies used in Dullah Laut except for gillnet fishing. From information I collected regarding 27 lift nets which were made and used in Sathean Village since 1983, I found that fishermen spent from Rp1–5.5m, and how much a fisherman spent depended on when he constructed the lift net and how big it was. The price of the lift net—including the net, planks, and nails—has increased over time. Of course the bigger the lift net, the more material needed and the more capital invested. In addition, extra capital was also needed for a boat to transport people and the catch between the lift net location and the village. The boat would cost a fisherman around Rp150 000 to Rp300 000. When I did my fieldwork, most of the fishermen equipped their boat with an outboard engine, requiring additional capital ranging from Rp750 000 to Rp2.5m.

There were several ways people got the money to construct and operate lift nets. The most important sources were savings from their previous work, borrowing from friends or moneylenders, and—a method that was becoming the favourite—a 'special' loan scheme operated by the fishing company. It is interesting that from the 22 fishermen I interviewed, there was only one who succeeded in accumulating his capital from his previous fishing activities, in this case gillnet fishing. There were another four fishermen who raised their capital from a combination of net fishing, sailing, and trading. I believe this indicates that fishing technologies before the lift net were not profitable enough to enable fishermen to accumulate the funds needed to upgrade the technology they were using. By contrast, borrowing from friends or money lenders was quite common and among those I interviewed, there were about 11 fishermen who acquired their capital in this way. Borrowing from friends was based on the close personal relationship between the borrower and the lender and economic calculations were not central to these arrangements. On the other hand, borrowing money from a money lender was purely economic and interest was always charged.

The special loan schemes introduced by the fishing companies were becoming popular when I did my fieldwork. The basic agreement involved in this scheme was that the company provided the fishermen with an amount of money to construct a lift net and if necessary, buy a boat and engine. In return, the fishermen were obligated to sell the main target of lift net fishing—anchovy—to the company. The company would deduct a certain amount from every purchase from the fishermen to repay the debt. The motivation for this scheme was purely economic; the company used the local fishermen to extract the resources and secure a continuous supply of the fish. Unlike other schemes, it was the lender (the fishing company) that actively looked for clients. Selection was based on two considerations—the borrowers' fishing ability and personal connections. The first consideration was to ensure that the invested capital resulted in a supply of fish. The second consideration sprang from the companies' need for people who could help them in handling paper work and dealing with local fishing and exporting officials. Given the importance of these connections, the chosen person was not necessarily a fisherman. However, whoever was chosen would usually ask skilled fishermen to operate his lift net.

It was quite a surprise to learn that despite the higher investment, lift net fishing did not generate higher income for its owner and operator compared with simpler and cheaper technologies used by Dullah Laut villagers. I found that the net monthly incomes of a fisherman working on two different lift nets were Rp159 755 and Rp104 480. In fact, these incomes were even lower than income from all other fishing technologies in Dullah Laut (see Table 8-1).

Marketing

Anchovy is a commodity that is sold both locally and exported. Regarding the anchovy taken by the lift nets involved in the incident I described earlier, they were all exported. In this particular context, there were three paths to getting the anchovy on the export market: direct marketing of the anchovy from the lift net owner to the exporting company; a marketing trough created by a small anchovy businessman who in turn would sell the fish to the exporting company; and another direct marketing from a local anchovy collector who also operated their own lift net and whose catch was sold directly to an export company.

The main export destination is Taiwan. As I mentioned earlier, the connection between lift net fishing and the international market was established in 1995 when a Taiwanese businessman started exporting anchovy to Taiwan from the Kei Islands. In that year, anchovy companies exported 370.5 tons with a total value of approximately Rp232.6m even though anchovy export businesses did not engage local middlemen.

The Political Economy of the Conflict

The clove season incident was triggered by the use of lift net fishing by non-Dullah Laut villagers who were provided the capital for their operation by a fishing company involved in buying and selling the catch to international markets (see Table 8-2). The lift nets were operated in Wada Iyuwahan, a territory belonging to Dullah Laut villagers who only used locally funded and relatively simple fishing technologies targeting locally marketed fish.

Table 8-2: Parties and issues relating to the conflict.

	Technique	Owner	Capital	Market
Village-based	Fish pot and trap, line and net	Dullah Laut villagers	Locally earned	Local
Non village-based	Lift net	Kei Islanders except for Dullah Laut villagers	Locally earned as well as outside company	Local and export

Source: Fieldwork research, 1996.

As with incidents I have discussed in other chapters, the 'clove season' incident involved multiple issues. However, I suggest that the main issues triggering this incident were economic and economic issues are an integral element of marine tenure.

The conflict between Dullah Laut and non-Dullah Laut fishermen came about through the adoption of new technology. Of course, this kind of conflict is not unique to the Kei Islands. In fact, conflict associated with the development of fishing technology is not unusual in any fishery around the world. For example, Matsuda and Kaneda (1984) found that technological development had been a significant factor in six out of what they called 'the seven greatest fisheries incidents in Japan'. Bavinck (2001) also found that using a trawler had caused serious conflict between the operator and fishermen using more traditional technologies who fished in the same fishing ground in India. In Indonesia, it was the bloody conflict between trawlers who operated in fishing grounds used by fishermen using smaller and much simpler technology, such as gill nets and small purse seine, that caused the Indonesian government to ban the use of trawlers in 1980 (Bailey 1986, 1997). I have also observed that technological developments created conflicts between fishermen in Bebalang Island, Demta, and Sathean villages in South Sulawesi, Irian Jaya, and Maluku respectively (Adhuri 1993).

There are two main questions that should be addressed when understanding how the conflicting economic interests between Dullah Laut fishermen and fishermen coming from outside gave rise to conflict. The first question is why were the outsiders driven to fish in Dullah Laut territory? The second question is why did Dullah Laut villagers object to the operation of lift nets and grouper fishermen in their territory? The following discussion will focus on answering these questions in particular.

A group of interrelating factors drove non-Dullah Laut fishermen to expand their fishing location outside of their territory and drew them to the Wada Iyuwahan fishing territory. The first factor was the spatial requirements of lift net technology. Although the actual size was no bigger than the stake trap, the operation of the lift net needed much more space because the light used in lift net operations attracts fish from a radius of 30–50 metres. Given this, it was assumed that the distance between operating lift nets should be around 60 meters. Therefore, as the number of lift nets increased, more fishing space was needed while the number of suitable fishing spots in their territory became limited.

The second factor was that the fish catch in the outsider's territory had decreased due to the increasing number of lift nets operating in their territorial waters as well as seasonal change. The territorial expansion of operation was an attempt to avoid further deterioration of the catch for individual fishermen. The seasonal change referred to the fluctuation of the targeted fish as a result of changing weather. The number of anchovy was reduced during the west monsoon when their territory is subject to strong wind, which contributed to the fishermen being forced to operate their lift nets elsewhere.

A third factor was the abundance of anchovy, which traditionally signal the presence of the trevally, in Wada Iyuwahan. When the fishermen were informed that the clove season was located in Wada Iyuwahan, they treated it as an invitation to move their lift net operations.

These fishermen were local people who were well acquainted with the concept of communal marine tenure, which meant they were aware of the traditional procedures used to gain access to fishing spots in Dullah Laut village. For the first lift net owner, this could have been done by asking permission from the village head of Dullah Laut. However, when the first lift net owner found that his father-in-law claimed rights to the territory, he decided that such a request was unnecessary—a view corroborated by Mr H when he claimed ownership of Dullah Laut by recalling the narrative of the Papuan War.

The last factor was the result of 'encouragement' from the companies that provided the capital for the lift nets. The importance of this encouragement was that it outweighed the perceived risks of operating 'illegally'. One such fishing company involved with the lift nets operated in Wada Iyuwahan. The scheme agreement was the same as it was in other areas—the fishermen were obliged to sell their catch to the company and the company would deduct the repayments from the purchase price. This scheme however, did not have a provision for fisherman to repay the loan if the operation of the lift net failed before the investment was fully repaid. As a result, the fishermen interpreted the company's encouragement as a sign that the company would take responsiblity for the risk faced by the fishermen. In this sense, there was nothing to lose for the fishermen. If the worst came to the worst, such as the destruction of their lift nets, it was not their fault but the fault of the company. It was also the company's loss, not theirs. On the other hand, if they succeeded they got to benefit from the 'illegal' operations.

From the company's point of view, the encouragement was necessary in order to maintain the continuity of fish supply so that they could meet the export demand and make a fast return on their investment. Of course it is likely that they took the risk of this encouragement into consideration since they were aware of the existence of traditional marine tenure practice. However, in their view the worst possible consequence would be eviction from Wada Iyuwahan (which was in fact, what eventually happened). More serious consequences such as physical abuse and lift net destruction—although perhaps 'legal' according to traditional practice—would be considered criminal acts under Indonesian law which would have made physical abuse or lift net destruction less likely and also opened the possibility for the company to sue Dullah Laut villagers for any damage caused to their equipment.

Now to answer the second question, which is why did the Dullah Laut villagers object to the operation of the lift nets in Wada Iyuwahan? Since the answer to this question lies in the practice of marine tenure, I would like to reiterate the basic tenure concepts. In Chapter Five, I described how the technology and level of exploitation defines the exclusivity of village sea territory. The question of who may make use of or own a territory involved two kinds of rights attached to the territory: use right and property right. It should be noted however, that the meaning of property right here does not mean 'the absolute possession of all rights or almost all rights by a single party' (Crocombe 1974: 8). In fact, the main difference between the use right and property right is the additional right to transfer the use right. This means that those who hold property rights on a territory are free to make use of the territory or extract available resources, and are also able to transfer their use right to other parties.

In Chapter Five, I also noted that property right and use right are held by particular social groups. In the case of Dullah Laut village, property right is the privilege of the origin group called *Ohoiroa Fauur*. Use right is held by those who have marriage connections with *Ohoiroa Fauur* members. Unless the property right holding unit grants permission, outsiders cannot extract resources from the territory. However, this exclusivity does not apply to outsiders if their level of exploitation is considered to be only for subsistence purposes and they have a good relationship with Dullah Laut villagers. The judgement on the latter was the privilege of the right holders of the territory.

Like the Morovo people in the Solomon Islands (Hviding 1989), the people of Dullah Laut have often said that everyone could fish in their sea territory as long as it was only for subsistence. How did the villagers of Dullah Laut make judgements about this 'subsistence' purpose? It is clear that 'subsistence' is not interpreted in a strict 'hand to mouth' sense. For example, fish pot, stake trap, and line fishing are all regarded as subsistence fishing technologies despite the fact that the catch is sold. Nor does the level of income appear to play an important role in defining subsistence operations. In fact, the data from the previous sections shows that the economic return for lift net owners tend to be lower than the return to those operating traditional technologies.

There are four factors that appear to be the most important when making judgements on whether a fishing operation is a subsistence operation or not. The first is the nature of the technology. Whereas the Morovo people consider technology to be separate to the issue of subsistence fishing, for Dullah Laut villagers it is a defining characteristic. The second factor involves an assessment of the level of exploitation. The third relates to the capital investment required, and the fourth concerns the nature of the market in which the catch is sold.

When I asked an informant why the lift net operation at Wada Iyuwahan triggered the incident, he referred to the first two of these factors: 'Imagine! A lift net, at least 10-by-10 metres in size, catches tons of fish in one haul compared with line fishing that only gets one fish in every catch. Is it comparable?' Technically, lift nets are bigger and their construction is more complex than local technologies operated by the villagers. The technical complexity is obvious if we note that the construction of a lift net requires a specialist while almost every fisherman can prepare fish pot, stake trap, lines and nets. Dullah Laut villagers also consider spears and locally made rubber 'guns' to be subsistence-oriented fishing gear.

Dullah Laut villagers considered lift nets to be much more exploitative than the technologies they used. Actually, it is difficult to compare the exploitation levels of lift net with the other fishing technologies because they target different types of fish and sell them in different units. However, it seemed that people measured the level of exploitation by comparing the maximum possible catch. In this sense, lift nets were proven to be more exploitative than line fishing, as demonstrated by the level of catch in the incident I described. Furthermore, the operation of the lift net also disturbed line fishing because they caught the fish that were in the water to attract the line fishermen's' target fish.

The capital investment required for lift net operations also supports the belief that this type of fishing does not fall in the category of subsistence fishing. Although in some cases net operations had larger investments, lift net operations require a large initial investment whereas net fishing operations can be carried out with the minimum size of net and gradually expanded. Lift net construction requires complete construction at the commencement of the operation.

The issue of marketing is also taken into consideration when defining subsistence activity. In this regard, fishing activities aimed at the local market as opposed to regional or export markets are considered to be at the level of subsistence. In this sense lift net fishing clearly fell into the non-subsistence category. In fact, it was the companies involved in export marketing that started the operation of lift nets in Dullah Laut territory. All but one of the lift nets—the first one, owned by a Butonese—were capitalised by a fishing company that bought the catch and exported it to Taiwan. This direct marketing from lift net fishermen to the exporting agencies could also be seen as by-passing the chain of marketing that supported local middlemen. Thus, from the point of view of the Dullah Laut villagers, the lift net operation was not only 'illegal' in relation to their marine tenure but also directly threatened the economy of their fishing activities.

Given that they were considered commercial operations, there were only two possible scenarios under which the lift net operations could have been permitted. If the lift net fishermen had use right over the territory, their activities would have been allowed. But, both the owners and the operators of the lift net were non-Dullah Laut villagers and they had no marriage connection with *Ohoiroa Fauur* members. Thus, they were excluded from any rights over the territory. Their attempt to gain use rights through the Ut Island villagers was also unsuccessful. This would only have been possible if the Ut villagers held property right over Wada Iyuwahan. In fact, this attempt led to an even more serious reaction from the Dullah Laut villagers. Mr A's claim that the location of the clove season—he called the location Metan Er—was the property of the Ut villagers particularly offended some of the *Ohoiroia Faur* leaders. This claim was taken to be a challenge to the territorial sovereignty of the Dullah Laut villagers. Thus, the lift net operation not only questioned use right but, more seriously, challenged their property right over the territory. The latter was considered to be a declaration of war by some of the *Ohoiroa Fauur* leaders and Dullah Laut villagers were prepared to enter into physical conflict with the Ut villagers. According to a traditional saying: 'people are willing to die to defend the boundaries of their island and land' (*umat her mat utin nuhu tanat …*).

The second scenario under which the operation could have been permitted was if the lift net owners had contacted *Ohoiroa Fauur* representatives and asked them for permission to fish in their territory. Under this scenario, a leasing agreement which would allow the transfer of use right from the original right-holding unit to a second party could have been drafted. A contractual fee (*ngasi*) would be involved under such an agreement.

Conclusion

It is obvious that the clove season incident was brought about by the technological developments that had occurred in the Kei Islands. Ultimately, it was the international anchovy market that stimulated the fishing company to operate in Kei Islands, and it was for the sake of the market that these companies funded fishermen in order to develop a greater and more secure supply of fish.

Has technological development degraded the practice of marine tenure? My answer to this question is no. What the incidents shows us is the opposite. When the lift net was used by outsiders operating in Dullah Laut territory, the people of Dullah Laut strengthened their traditional marine tenure practices. They did it by recalling narratives of origin to remind all parties involved of the sea boundaries and to declare their exclusive rights to their sea territory. Customary meetings involving as many customary leaders as possible were held to confirm every possible version of the narrative of origin. Once they were certain of the answers, they stood firm and exercised the necessary action to uphold their tradition. Interestingly, customary marine principle was also used by Ut leaders to make the descision to disregard protocols and lay claim to Dullah Laut marine territory. Origin narrative was chanted to lay claim of ownership, or use right at the very least. Under these set of circumstances, conflict was the fruit.

9. Marine Tenure and Precedence Contestation: A Village Destroyed[1]

This chapter will discuss a conflict between Sather and Tutrean villages on the northern coast of Kei Besar Island (Map 1-2). This conflict is very important for a comprehensive understanding of the problem of communal marine tenure in the Kei Islands. This is not only because the conflict was very cruel and had persisted through almost a century, but because the conflict questions the very foundation of traditional communal marine tenure.

By looking at the history of the conflict and examining the failure of attempted solutions initiated by the Dutch and Indonesian governments, this chapter will argue that even in the context of tradition, the concept and practice of communal marine tenure are problematic. This is because controls over territorial and political domains are two main points of contestation over precedence between different traditional segments of the community. Thus, although the triggering factor has been associated mostly with the sea territory, the core issue of the conflict between Tutrean and Sather villages was a traditional contestation over precedence between the *mel* and the *ren*.[2] For the noble of Tutrean, controlling the sea territory of Sather Village is a symbol of their precedence over the free people of Sather. For Sather villagers, having their own sea territory was a symbol of their freedom. Since tradition has taught them that fighting for their territory is a legitimate reason for 'war', conflict was inevitable.

When a Village was Burned to Ashes

In the beginning of this book, I described an incident involving the burning of Sather Village by Tutrean villagers in 1988. (See Plates 9-1 and 9-2 for images of the reconstruction) This is not an isolated conflict between the two villages and is one example from a limitless series of incidents that have occurred during almost a century of conflict. Conflict over coastal boundaries alone can be traced back as far as 1935. From that time on, particularly when the Trochus shell became economically valuable in the 1950s, rarely did a year go by without a fight. The climax was the incident I described.

---

1 A shorter Indonesian version of this chapter was published in Adhuri (1998b).

2 At the latter stage of the conflict, the *mel* broke up into two factions, which caused the conflict to then involve three different parties. Yet, it did not eliminate the hierarchical dimension of the conflict between the *mel* and the *ren*.

Plate 9-1: A Sather elder standing in front of the foundation of a destroyed house. A new house was built in the back of the foundation.

Source: Author's photograph.

Plate 9-2: A house (with accompanying clothing line) now resides on the foundation of a home that was destroyed during the 1988 Sather village fire.

Source: Author's photograph.

Ironically the destruction of Sather Village did not stop the 'war', as it is commonly called. Spears, stones and curses are still exchanged between the people of these villagers. For example, during a year-long research excursion in the Kei Islands from February 1996 until March 1997, I heard that four fights broke out between the two villages, all of them triggered by fishing activities conducted by either party in the disputed area.

Several attempts have been made to settle this conflict. On 20 February 1936, the Dutch representative arranged a meeting of the committee consisting of the Dutch representatives and prominent kings in the Kei Islands. The meeting, held in Weduar Village (Figure 9-1), discussed the conflict between Sather and Tutrean. At the end of the meeting, a conclusion concerning the boundaries of the two villages was reached and a legal decision was then issued.[3] However, at least one of the parties was not satisfied with this decision and the conflict continued.

A second attempt was made in 1939 when the Dutch cancelled the 1936 ruling and signed another decision establishing the boundaries between the two villages. As with the first decision, the second ruling was not strong enough to restrain both parties from violent behaviour. Another attempt was initiated by the Maluku Tenggara head of regency in 1990, two years after the village of Sather was burned to ashes. The head of regency—a Keiese— arranged a customary court in Elat on Kei Besar Island. After his attempt to resolve the conflict between the two villages failed, the head of regency urged the Sather villagers to bring the case to the Maluku Tenggara regency court in 1993.[4] This was the third attempt to solve the conflict. The Sather villagers brought the case to the court in Tual in 1995 who decided that Sather villagers and the descendants of Kapitan Liberth Rahantoknam both shared use right over various territories. However, all parties were not satisfied and they appealed the case to the provincial high court. In 1997, just before I finished my fieldwork, the decision from the high court was handed down in the regency court. I do not know how the villagers responded to this decision but judging from the distribution of the disputed sea territory in the decision, the Sather and Tutrean villagers will appeal to the Supreme Court and the conflict will continue.

The following discussions will examine the nature of the conflict by looking at its development over time. I will start by looking at the narratives of origin that describe relations between these two villages. As we will see, the narratives of origin were the starting point for conflict between the two villages. I will then discuss the conflict during the Dutch period examining the interventions of

---

3 Unfortunately, this legal decision was not available and no one knows what the outcome was.
4 Actually the head of sub-regency at Kei Besar had urged both villagers to bring the case to court before Sather village was burned.

the Dutch and the reactions of the Sather and Tutrean villagers toward these interventions. Finally, I will discuss the controversy surrounding the customary court that took place in 1990, the Maluku Tenggara Regency court hearing, and the Maluku provincial court hearing.

Narratives of Origin and the Conflict

Conflict between the Sather and Tutrean villages was present even in the narratives of origin. This is very obvious if we listen to the narratives told by the noble of Tutrean and the free people of Sather. Although the narratives from both parties address the same issue—the 'history' of the current Sather domain—the two versions are very different. The narrative told by the Tutrean villagers goes as follows:

> In the ancient time, Sather and Tutrean were two separate villages. Each was autonomous: each controlling its own territory and governing its own domain. At a particular time, conflict broke out between these two villagers. The conflict led to a war. Many people from Sather village were killed. Fearing for their lives, the rest of the Sather villagers fled to the island of Dullah in the Kei Kecil Archipelago. So, the Tutrean villagers won the war and, since there was no one left in Sather Village, [the] Tutrean claimed ownership over the territory. They called this territory as the 'land of victory' (*tanah kemenangan*).

This particular segment of the Tutrean narrative was used as the basic reference of their argument that the Tutrean territory covered two units of territory. The first unit was their own while the other unit was the 'land of victory'—the former Sather village territory. When they were asked about their sea territory boundaries, the Tutrean villagers would point to Yewukil which marked their coastal boundary to the north. This was the boundary between Tutrean and Weduar villages. In defining their boundary to the south, the Tutrean villagers would point to Wautkowar. This was the coastal boundary between the current Sather and Kilwat villages.

The Tutrean narrative also recounts the followings:

> Some time later, two groups of people came to the former Sather territory. These groups were seen by Yayat Rahabeat who reported their arrival to Tabal Tanlain, the leader of Tutrean village. Tabal Tanlain met them at the shore where they had landed. He asked these people who they were and their intentions. These people answered that they were *ren* Waer

Waw and *ren* Waer Rat, from Waer,[5] a village on the northern coast of Kei Besar. Their intention was to find a place where they could live. When these meeting[s] were concluded, Tabal Tanlain decided to allow them to live in a place named Ohoi Twu. In return these people were asked to take care of the 'land of victory' (*tanah kemenangan*). These people were the ancestors of the present Sather villagers.

In the course of their life, members of the Waer groups and their descendants often committed serious mistakes that were subject to severe punishment. For example, one of the mistakes involved some people from Ohoirenan who were taking a rest and having a meal on their way to Elar. Six people of the Waer, who were collecting seashells and fish in the coastal area, called them *suanggi*.[6] The people of Ohoirenan got angry and brought these six people to Ohoirenan. They reported to the head of their village that the six people had humiliated them. The head of Ohoirenan Village held a customary court that decided that these people had violated the customary rules. They were fined in the form of traditional wealth.[7] The head of Ohoirenan Village sent a courier to Tutrean to inform them about the case. Having been informed of what had happened; Tabal Tanlain paid the fine in the form of a gong, an antique canon (*lela*), and an elephant tusk. When the fine was paid, the six persons were sent back to their houses. Similar events directed to people from different villages occurred again and again. And the noble of Tutrean were forced to pay the fines.

For the noble of Tutrean, the above narrative fragment clearly shows that the current population of Sather is descended from free outsider people. The narrative implies that the *ren* of Sather are 'bought people' whereas the original meaning of *ren* means free people (see Chapter Four). This is because, according to the narrative, the ancestors of these people were those who were bailed out when the Tutrean noble paid their fines. It was a common belief that those who were bailed out became dependant or were 'owned' by the person who paid for them. Therefore, although the noble of Tutrean have never referred to the Sather villagers as the *iri*, they treated them as such because they believed that their ancestors bought them like slaves. These descriptions were used as the basic argument of the *mel* at Tutrean when they said that the *ren* of Sather had no political rights on their domain. According to the *mel*, it was clear that by definition, the *ren* were excluded from taking part in the political life of the

---

5   This narrative assumes that these people were free people driven away from their homeland. Thus, this narrative is not the same as the narrative that explains the history of social rank formation (see Chapter Four).
6   Van Hoëvell (1890: 127) noted that *suanggi* was a bad spirit in the shape a person who had magical power to cause disease and illness. Those who were proven to be *suanggi* were killed and their corpses thrown into the sea. Therefore, the accusation of being a *suanggi* was the worst accusation that could be made.
7   The traditional wealth consists of gong, antique cannon and plate, and gold ornaments of different kinds.

domain (see Chapter Four). In fact, the *ren* of Sather as dependant *ren*, were considered to be almost equivalent to the *iri*. Therefore at the very least, the Sather villagers had no right to have a village head from among their own people.

Regarding territory, the narrative also notes that the *ren* of Sather are very different to the *ren* who sprung from the land or from animals living on the land—that is, the *ren* who hold the title of lord of the land. Thus from the *mel* point of view, the *ren* of Sather have no special attachment to the place where they live. Therefore they are not entitled to posses the land and sea. They only live there because Tabal Tanlain was generous enough to allow their ancestors to live at Sather.

The *ren* of Sather reject this narrative, of course, and propose their own version instead. Their narrative says that their territory was part of a large kingdom called Tabab Yam Lim. The kingdom was led by Tabal Tamangil, who lived at the current Tamangil Nuhuten Village and divided his territory into five distinct territories. These are the current Tamangil, Kilwat, Sather, Tutrean, and Weduar Village territories. The first, the territory of the current Tamangil Village, was given to Rahanar, Ohoiner, and Badmuar.[8] The second, the current Weduar village territory, was given to Wowoa and Rahawarin. The third, the current Sather village territory, was given to Jamco and Jamlaai. The fourth, the Tutrean territory, was given to Safik. The last, Weduar territory, was handed over to Rahajaan, Limduan, and Rafo. Each of these people then became the lord of the land in each territory.

This narrative was used by the *ren* of Sather in opposition to the narrative of the *mel* of Tutrean. It was the basic reference point for their claim that their village was a separate territory from that of Tutrean and the other three villages of Tabab Yam Lim. Therefore, they argued that the Tutreans' claim over Sather territory had no 'historical' basis. The owners of Sather territory, Jamco and Jamlaai, were the lords of the land (see Plate 9-3).

Concerning the immigrants Waer Waw and Waer Rat, the Sather narrative states that they were accepted by the lord of the land of Sather, not by the Tabal Tanlain in Tutrean. Therefore their arrival had nothing to do with the *mel* in Tutrean. From the point of view of the *ren* in Sather, this again is proof that any claims that the Waer groups were the dependants of the Tutrean *mel* were false.

---

8   All the names mentioned as those who received the distribution of the territory were considered to be the original inhabitants of the villages.

Plate 9-3: An elder of Yamko *fam* showing an antique betel nut container which serves as proof of his authenticity as land lord of the Sather village.

Source: Author's photograph.

In brief, the narratives told by the Sather villagers justify two claims: first, that their village is a distinct domain and is independent from Tutrean, except that they were once both within the domain of Tabab Yamlim. Second, that the people of Sather are independent *ren* who do not belong to or under the control of another group of people. For the *ren* of Sather, these two claims demonstrate that they hold full rights over their territory as well as the right to govern their own domain.

The Dutch and the Conflict

The people of Maluku have experienced many changes since their first contact with the VOC and the Dutch colonisers back in 1605. The changes mostly arose from attempts by the VOC and the Dutch colonisers to achieve the economic and political goals of their occupation, such as the incorporation of local traditional social organisations into the Dutch political organisation. This was the colonisers' way of creating connections to the local people without making new structures that could be economically expensive and perhaps ineffective.

This strategy was designed not to change the local political structure, given that such changes could have created disturbances in the community which would have caused problems for the Dutch. However, this incorporation did lead to noticeable changes in the community. First, the Dutch government became involved in appointing traditional leaders and their treatment of these leaders was based on Dutch political assumptions rather than on the principles of traditional social organisation. Second, local people used the incorporation of their traditional structures into the Dutch political organisation in order to pursue their own political needs in the community. In Chapter Two, I discussed a case where a leader in Faan village used the Dutch government and the Catholic mission to appoint him as a king in opposition to the King of Tual. The following case concerning conflict between Sather and Tutrean villages is another example and in examing it, we will see that both villages manipulated Dutch involvement in their struggle.

The beginning of the conflict between Tutrean and Sather villagers, as people remember it, dates back to some time between 1910 and 1920 when the people of Sather began to demand that they be able to govern their own domain.[9] The people of Tutrean refused their proposal and it was decided that one of the leaders in Tutrean, Kapitan Liberth Rahantoknam, would be sent to live in Sather and be appointed leader of this village. Thus, although Sather became a separate village, they were still controlled by the noble of Tutrean.

---

9   It seems that prior to this period, Tutrean and Sather were organised as a single village with a single leader based in Tutrean. Thus, the narrative of ownership by the Tutrean nobles might hold some historical truth.

Sather villagers felt cheated by the decision. At that time, the leaders of Sather were divided into two groups. At a meeting which was mediated by nobles from Weduar and Ohoinangan villages, the leaders were asked to discuss whether they would accept the appointment of Kapitan Liberth Rahantoknam as their village leader. Each of these groups was told that the other group had accepted the proposal. Understanding that their fellow villagers in the other group had agreed, the Sather village leaders in both groups accepted the proposal.

The Dutch government issued a letter of appointment for Kapitan Liberth Rahantoknam as the traditional village head of Sather in 1917. This was good news for the Tutrean villagers because this appointment meant that the Dutch government accepted the 'truth' of their narrative of origin. This also led to the belief that the Tutrean nobles had won the contestation about the village leadership.

From the point of view of the people of Sather, Kapitan Liberth Rahatoknam's appointment was disadvantageous. First, it signalled that they had lost their contestation on the issue of the village leadership, and second, this appointment brought about a new notion regarding the position of traditional village head. Traditionally, at least from the noble's point of view, there was a distribution of power between the village leader or traditional village leader and the lord of the land. The former held power on political issues in the village, while the latter controlled issues of territory. The Dutch notion of village leadership did not distinguish between these two issues. In the Dutch conception, a traditional village head controlled both political and territorial issues. For example, the traditional village head was appointed as tax collector on land use and harvest—a position previously controlled by the lord of the land. This was interpreted by the Sather villagers as loss on the second battleground of their contestation, which was control over village territory.

Having been discredited by this new arrangement, the people of Sather moved to reject it. In 1927, they killed Kapitan Liberth Rahantoknam. This incident was even more controversial and profound because it was conducted by the Sather war commander who was considered to be representing the entire village. The killing was also carried out in Tual, the capital city of the region just as a big festival to welcome a Dutch official arriving from Ambon was occurring.

It is not clear whether the killing of the traditional village leader represented only a rejection of his appointment or whether it was a well-planned effort to remove the connection between Sather and Tutrean. Either way, the action was effective. The death of Kapitan Liberth Rahantoknam left the position of the traditional village head vacant which led to the appointment of the former traditional settlement head as the acting traditional village head. The dream of the Sather villagers came true. The former traditional settlement head was

Constantinus Domakubun, a Sather villager. With Constantinus Domakubun in the position of traditional village head, Sather was free from the domination of the noble in Tutrean. Sather became a distinct village with its own political domain. Half of the contestation had been won.

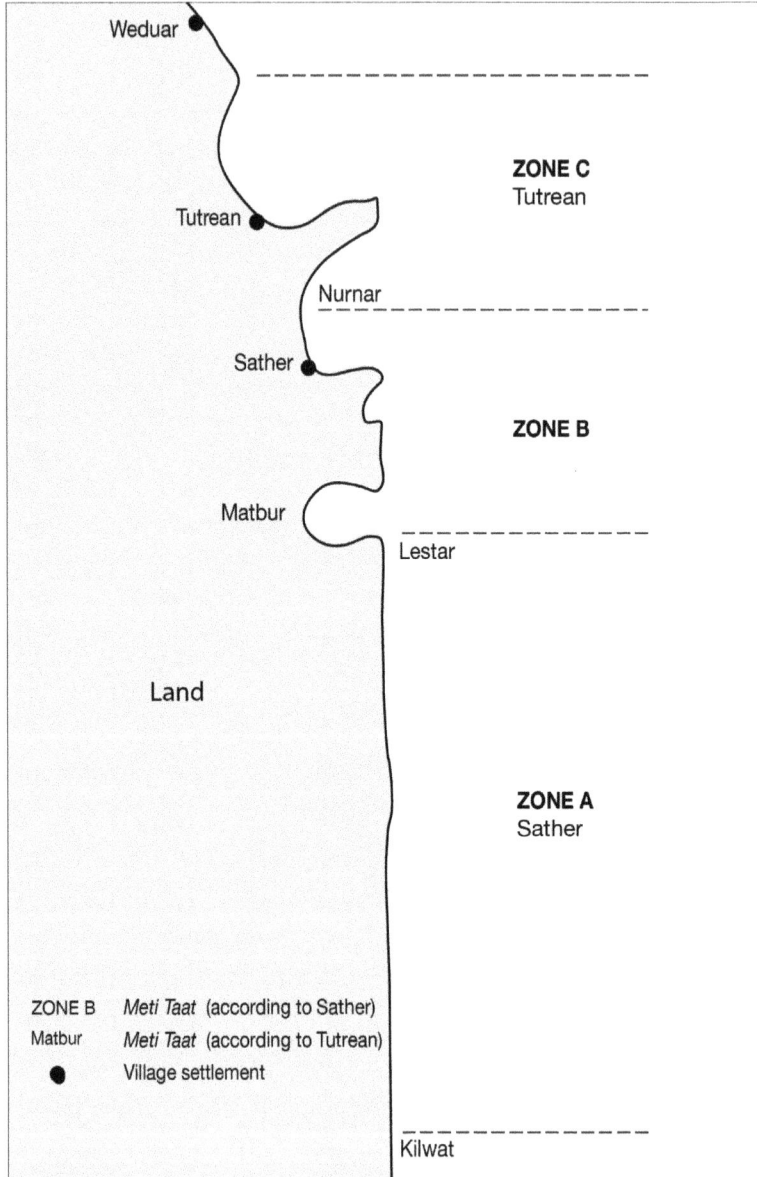

Figure 9-1: Tutrean and Sather coastal boundaries and sea territories, 1939.

Source: Adapted from Dutch Resumé (1939).

In 1935, the Sather villagers pursued the other half of their dispute with the Tutrean village by contesting the village's coastal boundaries. The dispute led to an open conflict resulting in the second Dutch involvement in 1936.[10]

In 1939, the Dutch again became involved in attempting to resolve the conflict. The 1936 legal decision was cancelled and a new one dated 11 September 1939 was issued. This legal decision stated that before the two villages reached an agreement on both the land and sea boundaries of their villages, they should comply with the following arrangement (Figure 9-1):

> The area marked Zone A on the map,[11] an area on the southern side of Matbur is for Sather. Zone C, the area on the northern side of Nurnar is for Tutrean. Zone B is a shared territory for both Sather and Tutrean with the following conditions. All private property in the house gardens and cultivated fields in Zone A and C that have been acknowledged should be considered as they are. The use of uncultivated land in Zone B is only permitted after the approval of the Dutch representative in Elat who will consider the matter based on the information from the village functionaries of both villages.

Rights over *meti Taat*[12] are taken from the two villages. A special committee consisted of the Dutch Domestic Administrator of Kei Islands, the Dutch representative in Elat, and a representative of the King of Fer who were responsible for arranging the use of this zone. The benefits taken from *meti Taat* will be distributed to both villages proportionally based on the size of population. Since the population of Tutrean is 473 and Sather has 346 people, the proportion will be 57.75 per cent and 42.25 per cent for Tutrean and Sather respectively. If both villages comply with this arrangement, an official will measure Zone B to identify the existing ownership over the garden. The committee will decide the ownership of the garden.

> It is not permitted for Sather to choose a traditional village head of their own, if they do not follow this arrangement seriously.

> It should be emphasised that these arrangements should be considered as a temporary settlement. It will be invalid once both villages reach a new agreement of their land and sea boundaries. The new agreement

---

10 Due to the Japanese invasion in World War II, the struggle for Indonesian independence, inclimate weather and other factors, many legal and historical documents have been lost or destroyed. I was unable to locate the 1936 legal decision so I can not explain what exactly the decision was, but neither party agreed with the decision and conflict between the two villagers continued.

11 I did not find the original map related to the legal decision. Therefore, the map produced to clarify this legal decision was based on the contemporary information.

12 The location of *meti Taat* is disputed. Sather villagers refer to zone B, while Tutrean villagers point to the coastal area between Sarwarin and Lestar.

can be strengthened with a legal formal decision by the *Grooten van Hoofden* (translated and adapted from the Indonesian translation of the legal decision, legalised by the Maluku Tenggara Regency Court).

It is interesting that this legal decision not only concerns the boundaries of the two villages, but also the internal politics of Sather village. In terms of sea boundaries, the decision verified the assumption that as a distinct village, Sather had its own territory. Regarding the village leadership however, the legal decision reverted to the environment that existed prior to the killing of Kapitan Liberth Rahantoknam in 1927. Although the legal decision does not explicitly state that if Sather did not comply with the new arrangement it would be under the leadership of the noble of Tutrean, this is the clear implication. Indeed, this became the reality when Gotlief Rahantoknam, a son of Kapitan Liberth Rahantoknam, was installed as traditional village leader in Sather in 1942.

As with previous interventions by the Dutch, this one did not satisfy the Sather villagers either. In fact, they considered the legal decision to be that of the Tutrean noble and legalised by the Dutch government. All of the decisions were made by prominent leaders in Kei Kecil and Kei Besar who were noble and who probably had affinal relationships with the noble of Tutrean. Thus, from the perspective of the Sather noble, the decision was unlikely to be fair.

The Indonesian State and the Conflict

The incorporation of the Kei Islands and all of Southeast Maluku into the modern Indonesian state also led to many changes in rural communities. However, the most rapid changes occurred during the New Order Regime (1966–98) when former President Soeharto carried out various economic development projects based on centralisation and homogenisation. Centralisation refers to government policies that were mostly crafted by central government agencies based in Jakarta. Homogenisation refers to government policies that were applied indiscriminately in all regions of Indonesia and paid little attention to the heterogeneity of Indonesian communities.

These policies introduced new structures to local communities. For example, state policy introduced the concept that every citizen is the same before the law. For those communities that have relatively strict systems of social stratification— such as in the Kei Islands—this is new and contrary to their traditional norms. The Indonesian government also formally replaced the traditional village with the modern village, and state courts with traditional courts.[13]

---

13 See Chapters Two and Six where the relationships between traditional and modern structures are discussed in detail.

The policies also provided an additional new context to the conflict between Sather and Tutrean. For the Sather villagers, the history of the conflict had taught them that tradition had almost always marginalised them given that traditional institutions were mostly considered to be the privilege of the noble. The involvement of the Dutch was also seen to favour the noble because the Dutch relied on traditional institutions to pursue their interests and actively made use of the influence of the noble, adopting them as collaborators. So for the Sather villagers, the application of modern state norms and institutions that did not discriminate between the noble and free people created the hope that a new door had opened which might lead to victory in their conflict with Tutrean.

Interestingly, the noble did not view the introduction of these new norms and institutions as a threat. For the noble, the new arrangements were not regarded as replacing tradition but enriching it. Since newly introduced state 'traditions' mostly dealt with power and resource distribution, they believed that this 'new tradition' would operate to their advantage.

I mentioned earlier that Gotlief Rahantoknam, son of the first traditional village head Kapitan Liberth Rahantoknam, was appointed the traditional village leader of Sather in 1942. Interestingly, during his leadership Gotlief Rahantoknam seemed to distance himself from the noble of Tutrean. In fact, according to some Tutrean informants, Gotlief's acts were in accordance with the will of Sather villagers. For example, he allowed some Sather villagers to open new gardens and cut trees in the disputed territory without any consultations with Tutrean. For the noble of Tutrean, these actions amounted to Gotlief's betrayal of his position as a representative of Tutrean nobility. This period was the beginning of the split between the Tutrean noble and the descendants of Kapitan Liberth Rahantoknam.

The Sather villagers were happy with what Gotlief did as traditional village leader. This was the behavior they had been hoping for since Kapitan Liberth Rahantoknam's leadership. A descendant of Sather's war commander even told me that if Kapitan Liberth Rahantoknam had governed as his son did, he would not have been murdered. In the beginning the villagers resisted Gotlief's appointment, but when he began to serve the people well, they changed their mind and supported his leadership.[14]

Gotlief was the first traditional village head in the period of the Indonesian government. In 1957, the Indonesian official in Elat issued a letter appointing him the traditional village head of Sather Village. Sather villagers supported

---

14   The people of Sather believed that Gotlief Rahantoknam's actions indicated he had adopted the *mel*'s perspective of his leadership. This understanding was explicitly mentioned in their legal debate at a session in the Southeastern Maluku Regency Court. They argued that the appointment of Rahantoknam as traditional village leader of Sather not only meant that Rahantoknam lived at Sather, but also that he socialised as a *mel*.

his appointment and in fact, wrote a letter to an Indonesian official at a similar level to the subdistrict head in Elat that nominated Gotlief for the position. The letter stated that Gotlief Rahantoknam was the descendant of the first traditonal village head, thus the proper candidate from the perspective of tradition. Gotlief was also considered to be a man who had good relations with villagers and others, was capable of leading people in development programs, and was prepared to take responsibility for dealing with the problems of land and sea boundaries between Sather and Tutrean.

This time, it was the Tutrean villagers who protested. In 1959, a petition signed by 13 noble leaders in Tutrean was sent to the Indonesian government representative in Elat. The petition explained their objections to the appointment of Gotlief Rahantoknam and demanded that the Indonesian government freeze the position of the traditional village head in Sather until they had settled the boundaries between the two villages as per the Dutch legal decision of 1939. The letter also said:

> The position of Gotlief Rahantoknam as the traditional village head of Sather does not mean that he is an original Sather villager, he only fills the vacuum of Sather village organisation, replacing his dead father who was a living historical fact [that the village organisation of Sather] is from, by and for Tutrean [the traditional village head] represents the right of Tutrean over the Sather village and its people for both internal and external matters. [This explanation] means that [the people of Tutrean] cancel their mandate to Gotlief Rahantoknam to represent their interest at Sather. (translated from a letter signed by 13 members of the committee of leaders in a village (*saniri*) 10 July 1959).

According to the statement above, the noble of Tutrean considered Gotlief Rahantoknam to have acted more as a Sather villager than a Tutrean noble. Therefore, he did not represent the interests of Tutrean any more.

These circumstances might lead us to conclude that conflict between the noble and the free people was not relevant anymore because Rahantoknam and his descendants who are noble joined the free people of Sather. However, this is not the case and this becomes evident if we follow the development of the conflict. In fact what happened is that the conflict became even more complex because this situation marked the beginning of the split within the noble group. Previously, the conflict was only between the free people of Sather and the noble of Tutrean but at this stage the noble came to be divided into two groups— the noble of Tutrean and the descendants of Kapitan Liberth Rahantoknam.[15]

---

15 There were Rahantoknam *fam* members at Tutrean as well, but they were on the side of the Tutrean noble.

Therefore, the conflict eventually came to be between three parties, and while the conflict between the two noble factions was significant, the longstanding conflict between the noble and the free people continued.

Since Gotlief Rahantoknam retired in the early-1960s,[16] no one among his descendants was interested in taking his position, but neither did they allow any of the other Sather villagers to take the position. Interestingly, the Maluku Tenggara regency leaders supported their objections most of the time. Therefore, from Gotlief's retirement until I finished my fieldwork in 1997, Sather Village was for the most part without a formal leader. I was unable to locate any information on who organised the village from the 1960s until the implementation of the *Village Law No. 5, 1979*, but after the village law was implemented, the village was mostly led by an administrative caretaker from the Kei Besar sub-regency office.

The above account indicates that despite the popularity of Gotlief's leadership, the Rahantoknam *fam* was not on the side of the Sather villagers. Their actions in preventing Sather villagers from taking over the leadership was made easier because Gotlief's son (Mr FL) was the leader of the 'village government' division at Maluku Tenggara Regency Office, which was responsible for village head elections. He told me that in the course of his leadership in the division, he had cancelled the village head election at Sather twice. The reason was that no Sather villagers had consulted him on the matter as they should have because he was a descendant of the first traditional village head. This is one example that illustrates the amount of control the nobles had over the local bureaucracy.

Furthermore, if we check the leading positions in the local regency and sub-regency government offices, we find that these positions are in the hands of the noble or outsiders who because of their positions, become colleagues of the noble. This has not happened by chance and in fact there has been a continuous effort by the noble to maintain the status quo. To give an example, in the 1980s, the governor of Maluku appointed the sub-regency head of Kei Besar, who happened to be a free person. As a newly appointed official, the sub-regency head arranged a meeting in his office to introduce himself to all of the village heads under his control. When the time came, not one of the village heads turned up. The reason for their absence was, as some informants told me, because he was a free person. 'How can a free person lead nobles?' an informant told me, quoting the reason uttered by a village head. Another informant explained, 'If he needs us, it is him that should come, not expecting us to see him. That is not the way it should be'. This incident was repeated several times before the

---

16 I could not verify whether his retirement was because of the petition of the *mel* of Tutrean or for other reasons.

sub-regency head gave up. Within six months of his appointment, he'd signed a letter of resignation and sent it to the Governor of Maluku Province. The Governor transferred him to another place outside the Kei Islands.

In light of the political circumstances discussed, I will now examine the issue of sea boundaries. From the petition signed by the Tutrean leaders, it was evident that conflict over sea boundaries was still an issue in the 1950s, and conflicts in the decades that followed resulted in the intervention of the sub-regency head of Kei Besar in 1987. On 20 July 1987, the sub-regency head invited the leaders of the two villages to discuss a possible solution to the dispute. The group agreed that the conflict would be brought before the court. Sather villagers would lay claim to the territory while the Tutrean villagers would be the defendants. However, this agreement was not implemented due to the burning of Sather Village in April 1988 which was detailed in the first page of this monograph.

As the conflict developed into a 'war', the regency head of Maluku Tenggara regency was prompted to intervene. Interestingly, he still tried to deal with the conflict in terms of tradition. The regency head via the head of the social and political division in his office, requested the head of the sub-regency in Elat to arrange a customary court which was held from 22 to 27 January 1990. The customary court was led by a committee consisting of prominent leaders from both the nine group and the five group, most of whom were kings (see Chapter Two). Several hearings were held to question the representatives from both villages and examine evidence. Finally, on 27 January 1990, after several hearings and a field trip to the disputed coastal region, the committee reached the following decisions (detailed in Figure 9-2):[17]

1. The boundary between Tutrean and Weduar up to Sarwarin is controlled by Tutrean.

2. From Lestar to the boundary between Sather and Kilwat is controlled by descendants of Kapitan Liberth Rahantoknam and Sather.

3. From Sarwarin up to Lestar is controlled by descendants of Kapitan Liberth Rahantoknam (Customary Court Decision, Elat, 27 January 1990).

---

17   The regency head who initiated the customary court strengthened the previous customary court decision with his decree No. 116/KDS/1990, dated 5 April 1990.

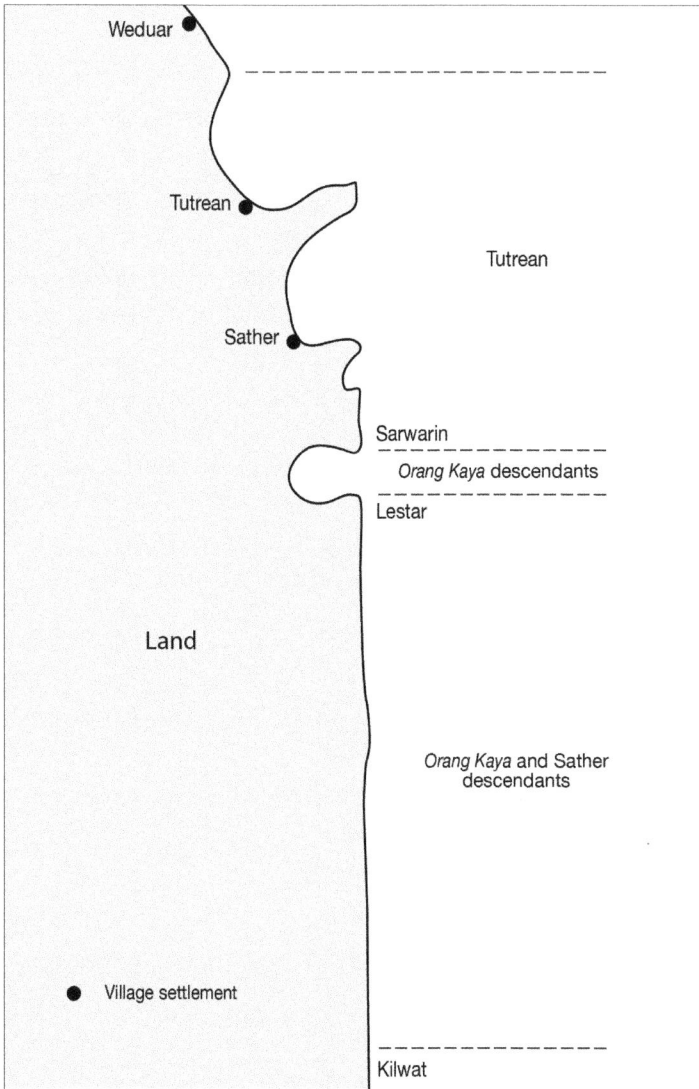

Figure 9-2: Tutrean and Sather coastal boundaries and sea territories, 1990.

Source: Adapted from Customary Court Descision (27 January 1990).

These decisions were based on the following considerations:[18]

1. That the main issue of the conflict was the location of *meti Taat*. According to Sather representatives, the location of *meti Taat* was from Nurnar to Lestar while Sather representatives pointed to a location from Sarwarin to Lestar.

2. Tutrean representatives recognised the right of Kapitan Liberth Rahantoknam's descendants by explaining that when leaders of Tutrean requested Kapitan Liberth Rahantoknam to live and lead Sather village, they gave him an area from a point called Year Karwin to Lestar within *meti Taat*. Therefore, they believed that Kapitan Liberth Rahantoknam's descendants owned this area.

3. Since both villagers acknowledged that Kapitan Liberth Rahantoknam was the traditional village head of Sather,[19] the Tutrean sea territory boundaries that formerly bordered with Weduar on the north side and Kilwat on the south side changed. [This was taken to mean that because Sather village had a traditional village head, they had their own territory.]

The decision by the customary court was disputed. Sather villagers were the most upset with the decision and if we examine the basic arguments of the decision, it is possible to understand why. The last two arguments were consistent with the Tutrean narrative that Sather Village was once under their control. From the perspective of the Sather villagers, the acceptance of the Tutrean version misled the committee in deciding the boundaries between the two villages as well as the location of *Meti Taat*. The effect of the decision was that Sather Village was not given control of any part of the territory.

In reality, Sather villagers had refused to attend the customary court. When the regency head of Kei Besar Island informed them of his plan to organise a customary court, they objected. In a letter dated 18 February 1990 and directed to the head of the Maluku Tenggara regency, the leaders of Sather explained that their objections were based on their desire to see those who burned their houses brought to justice.[20] They also informed the regency head that soldiers were pressuring them to attend the customary court by coming to their homes on the morning of the first day of the court sitting. In accordance with their agreement with the Tutrean villagers which had been arranged by the sub-regency head before the incident, the Sather villagers appointed a lawyer to

---

18   The considerations were not expressed in the exact way as done in this list, but because I am attempting to analyse the issues, I have paraphrased what is stated in the decision letter.

19   The committee of customary court had arranged for both villages to sign a letter stating that they accepted that Kapitan Liberth Rahantoknam was the traditional village head of Sather.

20   The burning of the village was brought to a criminal court. The regency court judge who handled the case sentenced 16 out of the 21 suspects to two years in jail. However, they appealed to the high court in Ambon and the Supreme Court in Jakarta. They were freed on appeal until the Supreme Court on 13 February 1996 decided to support the criminal court's decision.

bring their boundary dispute case to the regency court. I believe they must have known that the outcome of the customary court would be in favour of the Tutrean noble.

Before I continue discussing the regency head's response to the Sather villager's objection, I will discuss the links between the customary court and the political circumstances of the conflict. I mentioned earlier that the appointment of Gotlief Rahantoknam marked the beginning of the split between the noble of Tutrean and the descendants of the traditional village head. The decision of the customary court—led by a good friend of Gotlief's son— to allocate a section of sea territory to Kapitan Liberth Rahantoknam's descendants illustrates the split. In fact, the customary court's decision was used to gain an even greater portion of the disputed sea territory when Rahantoknam's descendants brought the case to court.

The case was brought to Maluku Tenggara Regency Court in October 1993. Sather villagers initiated the case claiming that Tutrean villagers[21] had violated their sea territory. Later on, the descendant of Kapitan Liberth Rahantoknam lodged another claim that as the descendants of the first traditional village head, they had the right to the disputed sea territory. Origin narratives, written documents, and witness evidence were discussed in the various hearings of the case. Finally after almost 18 months, the judges handed down their decision on 19 April 1995. As illustrated in Figure 9-3, it was ruled that:

> 1. Tutrean villagers have the right over the sea territory between Ohoimel and the boundaries with Weduar sea territory.

> 2. The descendants of Kapitan Liberth Rahantoknam controlled the territory between Sarwarin and Lestar.

> 3. Sather villagers and the descendants of Kapitan Liberth Rahantoknam shared use right over the territory between Ohomel and Sarwarin as well as from Lestar to the village sea boundary with Kilwat (translated from Maluku Tengara Regency Court Decision, 19 April 1995, No.20/ pdt.G/1993 /PN.TL).

---

21   Actually, some other parties were mentioned in the legal document prepared by Sather villagers but in the court hearings they were not discussed.

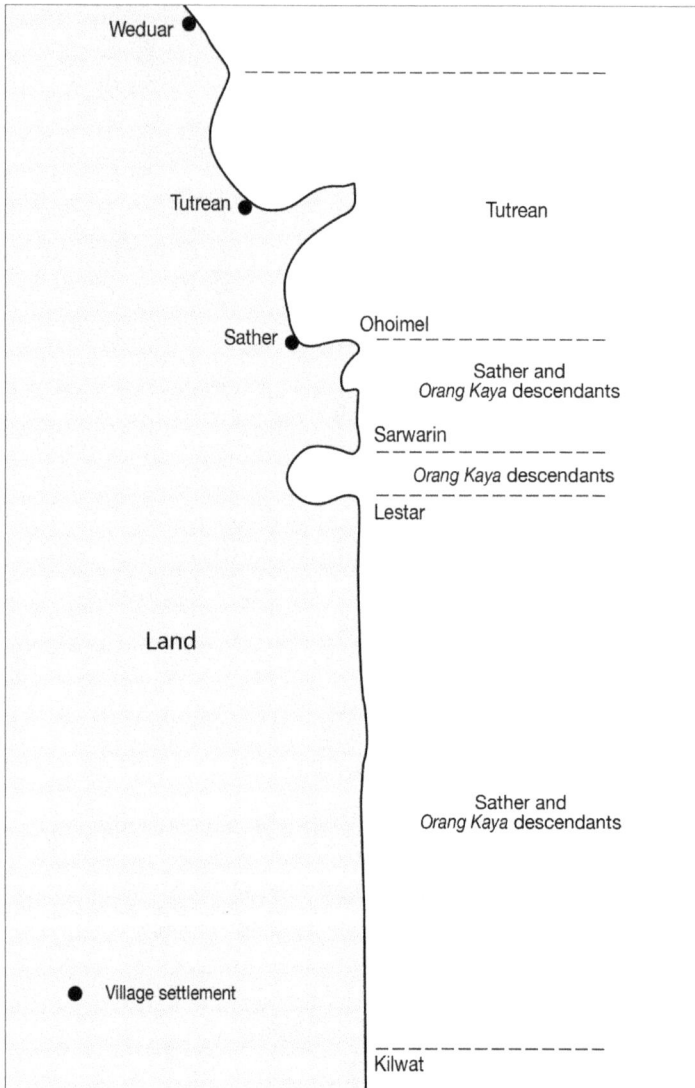

Figure 9-3: Tutrean and Sather coastal boundaries and sea territories, 1995.

Source: Adapted from Regency Court Descision (15 April 1995).

Like the previous decisions, this court decision was also rejected. The three parties appealed the case to the hight court in Ambon on May of the same year. In August 1996, the High Court in Ambon cancelled the regency court decisions and stated the following (see Figure 9-4):

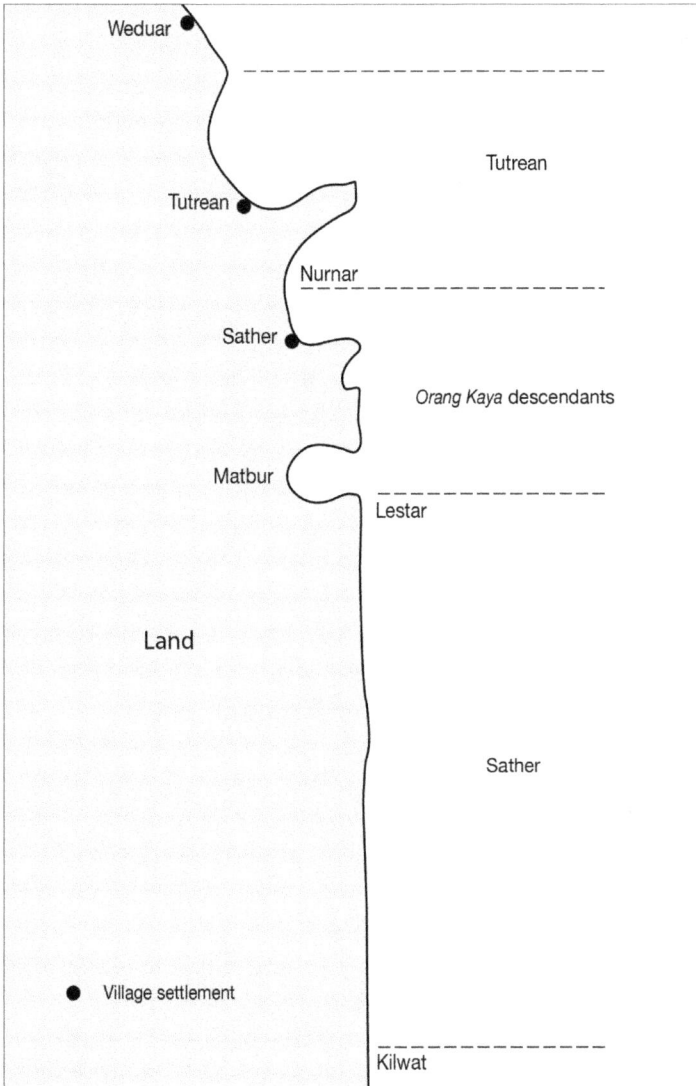

Figure 9-4: Tutrean and Sather coastal boundaries and sea territories, 1996.

Source: Adapted from High Court Descision (5 September 1996).

That Sather villagers owned the sea territory from Lestar to the sea border with Kilwat, and

> That *meti* Taat (from Nurnar to Matbur) was owned by Kapitan Liberth Rahantoknam's descendants as an inheritance (author translation from Maluku Province High Court Decision 5 September 1996).

I do not know how the three parties responded to this decision because when I finished my fieldwork, the regency court had just received the decision. None of the parties involved were aware of it. However, judging from the distribution of the territory, my guess is that Sather and Tutrean villagers would not be satisfied. This is because from their perspective, the distribution of the sea territory was worse than that of the disputed Dutch legal decision in 1939. The legal decision of 1939 decided that Zone B was under the control of a special committee with the harvest distributed proportionally to both villages, but the High Court decision ruled that the territory called Zone B belonged to the Rahantoknam descendants. Therefore, it was likely that the Sather free people and the Tutrean noble would appeal to the Supreme Court. The conflict, both in terms of legal pursuit and practical resource use on the disputed sea territory, was likely to continue.

Conclusion

The conflict between Tutrean and Sather villages demonstrates that the concepts and practice of marine tenure are embedded in the social structure of the community. The question of who owned what had a lot to do with the perceived owner's position in the community. In particular, the 'who' being referred to here are the social entities—the noble and free people. If the question of communal ownership was directed to the noble, they would respond that the free people were excluded from ownership.[22] The free people had a different perspective of communal property and would argue that it was the free people who were the lord of the land of their territory. We would also likely find different definitions of communal ownership if we asked different factions within the noble, which is demonstrated by the conflict between Kapitan Liberth Rahantoknam's descendants and the noble of Tutrean (see also Chapter Six).

In addition, the embeddedness of marine tenure in the social structure of the community is also illustrated by its inseparability from the political issues of the domain. If the political domain was not in the hands of the noble, questions about control of the territory would be answered differently. I believe that was why the first movement of the free people at Sather was towards gaining a traditional village head of their own. This movement was their political strategy to demonstrate that their village was a distinct and independent domain. If they had succeeded in removing the connection with the noble of Tutrean, their ability to pursue their own approach to communal property would have been enhanced.

---

22 I should note that parts of a territory are distributed to *fam* or families for ownership. Therefore, the term communal property refers to areas that are not subject to *fam* or family ownership.

In conclusion, I should also say that communal sea and land territory, as well as control over political domains were fields of contestation between the noble and the free people over precedence. On one hand, the noble insisted on defining their relationship with the free people as hierarchical, which was why they insisted on control over both the political and communal property domains. On the other hand, the free people assumed that their relationship with the nobles was a form of precedence—the noble controlled the domain of politics and the free people controlled the territorial domain. Even if they were both in the same domain, from the free people's point of view, there should have been a distribution of power. Given that they were actually in different domains, the free people should have been completely independent from the noble.

Finally, these conflicts revolve around tradition. The division of the Kei people into three social strata is a tradition. It is tradition that puts territorial and political domains as indexes of social status. And it is tradition that is the root of the Kei people's belief that no discussion of tradition is valid unless it concerns the narrative of origin. Adherence to tradition will perpetuate the conflict between Tutrean and Sather because of their conflicting narrative of origin.

10. Concluding Remarks

All around the world, from the coldest arctic regions to the warmest tropical seas, there is a crisis in the world's fisheries. Quite simply, there are too many people chasing too few fish. ... [T]hroughout the 1970's the world's per capita fish production actually declined. Correspondingly, the catch per unit of fishing effort and the catch per dollar invested in the fisheries also steadily declined (McGoodwin 1990: 1).

McGoodwin's findings should have come as no surprise. More than two decades before, Hardin (1968) had warned that this crisis might occur. However, the crisis is still disturbing because since Hardin's predictions, academics and resource managers from a range of disciplines have been trying to find solutions to avoid the crisis, and McGoodwin's study shows that they have not yet found an effective strategy. We might then pose the question: What is wrong with the discourse on fisheries management?

Following Hardin (1968) who argued that 'free for all' common property lay at the core of the problem of the discussion of fisheries management, many have come to agree that property rights are an essential element for creating sustainable and socially-just resource management systems. However, there is considerable debate over which property rights are the most suitable for fisheries management. Early on, private property and sole ownership, together with 'arrangements that create coercion' (Hardin 1968: 26) were proposed as the best solutions for coping with the tragedy of the commons. Nevertheless, evaluation of these solutions, which place government as a key element in resource management systems, has exposed many problems. These problems are associated with theoretical inadequacies (such as improper assumptions), implementation difficulties (such as market imperfections), and limits on government ability to effectively organise resource management.

Criticism of Hardinian thinking has led to increased discussion of traditional communal marine tenure which attempt to correct some of the misconceptions about human behaviour that give support to the 'tragedy of the commons' theory. For example, the assumption that people tend to maximise their self-interest in exploiting the commons is corrected by pointing out that the existence of communal property rights and associated regulations form the basis for cooperative action in resource use. Observations on other aspects of traditional marine resource management, such as traditional resource knowledge, alternative worldviews about the relationship between humans and nature, and social cohesiveness point to practices that are more effective and efficient than the solutions proposed by Hardin and his followers. The alternative studies convinced scholars and fisheries managers that communal marine tenure was the best way to manage marine resources sustainably and equitably. As a result, they suggested that governments should formally legalise the practice of traditional marine tenure and argue that the government's legal acknowledgement and support is needed in order to stop the decay of traditional tenure practices caused by the introduction and intensification of the market economy and 'modern' bureaucracy.

In Maluku the discourse on *sasi,* which as previously defined is the system of beliefs, rules and rituals pertaining to temporal prohibitions on the use of a particular resource or territory, has been a central element in the advocacy of community-based marine resource management. Early stages of the discourse on *sasi* promoted the view that it was inherently a sustainable resource management system and provided for a fair distribution of marine resources. More recent analyses note that the practice of *sasi* was based more on political and economic interests than sustainable and socially-just resource use. Since the practice of *sasi* involves multiple parties with various political and economics interests, the application of the practice differs from one party to another (Pannell 1997). These re-examinations of *sasi* have highlighted the more 'romantic' arguments developed at an earlier stage.

While I agree with these more recent views on *sasi*, this discussion has neglected an important aspect which I consider to be the main issue in the discourse on marine resource management: the issue of property rights—an issue originally put forward by Hardin. Although concepts related to sea ownership such as *petuanan laut* and *meti* or even communal marine tenure are sometimes mentioned, there is no discussion of the meaning and importance of these concepts for the community or theory of marine resource management. This book addresses this gap by providing a critical examination of communal marine tenure, and challenging the arguments put forward by those who are assured that communal property rights provide the best basis for marine resource management. At the same time, it is also a commentary on the discourse about traditional marine resource management in Maluku which has tended to focus mainly on *sasi*.

The core issue discussed in this book is the politics of marine tenure. Four conflicts have been analysed to see how 'politics' affects marine tenure in Kei Islands. Three of these conflicts broke out in Dullah Laut Village during 1996–97, namely: (1) the raid on an illegal fishing company; (2) catching the cyanide fishers; and (3) the clove season incident. The fourth conflict involves the Sather and Tutrean villagers and has been ongoing for more than fifty years. While all four conflicts pertain to marine tenure including the questioning of boundaries and distribution of rights over sea territory and resources, issues involved in these conflicts were various and complex. The following is a condensed version of my analyses of these conflicts.

1. Refering to the raid on the fishing company, I argue that this conflict was strongly influenced by the political jostling between three political parties over the head of the village position. The three parties involved were the modern village leader, the descendant of the traditional village head, and the head of the Christian settlement. To them, traditional marine tenure was critical to winning the village leadership and thus controlling marine tenure was seen as a demonstration of village leadership potential. The fact that these three political leaders in one way or the other were involved in leasing use rights for their cyanide fishing business is proof that resource sustainability and social justice were not their concern.

2. My discussion of the second conflict, triggered by the operation of outsider cyanide fishermen in village sea territory, focuses on the politics of legality. In this regard, I suggest that there is room to argue that formal law, regulation, and policy do accommodate traditional marine tenure. In fact, it was the rules and procedures of traditional marine tenure that were used to deal with the conflict through the power and interest of military and government officers. Traditional marine tenure was used by military and government agencies to cover up their involvement in the illegal business of cyanide fishing. The village leader, who tried to bring the case to the attention of formal institutions, was powerless against them. This case not only counters the popular discourse that traditional marine tenure is not aknowledged in formal Indonesian law, regulation and policies, but also shows that marine tenure and its relevance to resource management is much more complex. It demonstrates that formal acknowledgement alone does not make traditional marine tenure a self-contained institution. In practice, the power structure between traditional right holder units and external agencies determined the implementation of marine tenure. The case also demonstrated that formal legality did not guarantee the effectiveness of communal marine tenure in terms of resource sustainability and socially fair distribution.

3. The clove season incident looks at the relationships between market forces and the institution and practice of customary marine tenure. Through the analysis of this incident, I argue that the market economy does not always necessarily degrade traditional marine tenure. My evaluation of the impact of the international trade of frozen anchovy shows that people have been prompted to strengthen traditional marine tenure in order to exclude others from fishing in their territory. While I do not believe that forms of marine tenure 'appear as if from nowhere' (Hviding 1989: 5–6), there is no doubt that people manipulate and revive traditions in order to secure access to newly valuable resources. To express it differently, as the value of a resource increases, people strengthen their claims on communal marine tenure and use them to play the politics of exclusion.

4. My analysis of the Tutrean and Sather villages' conflict generates two conclusions. First, even in terms of tradition, communal marine tenure is a subject of contestation. This is clear from the conflicting claims about sea boundaries made by both Tutrean and Sather villagers that are in reference to narratives of origin—a main source of all discourse and practices of tradition. Furthermore, through various stages of dealing with the conflict, the government involved traditional institutions and leaders were always consulted. Yet, the conflict that had lasted more than half a century did not come to an end. The second conclusion notes that the conflict over marine boundaries between these villages was not only about marine tenure per se, but about precedence contestation between the noble of Tutrean and the free people of Sather. For the noble of Tutrean, controlling the sea territory was a traditional symbol of their nobility in relation to the free people of Sather. For the Sather villagers, it was a symbol of their independence, of being free people. Thus, it was very clear that communal marine tenure for these villages involved the social relations between the two in the social structure of the community.

If I were to define or identify communal marine tenure in Kei Islands, I would describe it as the following: communal marine tenure in the Kei Islands differs from the systems described in the current literature. For example, in the Kei Islands, not all of the rights over a sea territory are distributed equally to all members of the right-holding unit. It is only the use right that is distributed equally to all members of the community (Chapters Five and Six). In fact, even those who have an affinal relation are granted this right. By contrast, the property right is only in the hands of the core members, in this case the origin *fam* of the community (Chapters Four and Six).

Another important feature of communal marine tenure in the Kei Islands is its fluidity. When I note that property right over sea territory is in the hands of the origin *fam*, the question of what is the definition of this group arises. In the case of Dullah Laut Village, they are from the noble origin *fam*, as demonstrated by the decision-making over property right claims being made by the political leaders in the village, and the nobles holding exclusive authority over political issues. Nevertheless, if we question who are the origin nobles that really hold the property right, we will not find an answer that is agreed upon by all origin *fam* because there is disagreement between origin *fam* members over the issue of leadership. In Chapter Six for example, I discuss a conflict between the modern village leader and the descendant of the traditional leader over the issue of who held the right to allow a Taiwanese businessman to fish in the village sea territory. In the case of the conflict between Sather and Tutrean villages (Chapter Nine), the definition of the right-holding unit is even more contentious with long-term and violent disagreement between the nobles and the free people over who holds power in territorial and political domains.

The fluidity of communal marine tenure practise exists not only in terms of the right holding unit but also in relation to sea boundaries. In all villages discussed—Dullah Laut, Tutrean, Sather, and many other villages that I have not detailed—the sea boundaries between one village and another have been a significant source of conflict (see Chapters Four, Five, and Eight). This is a clear indication that there are no fixed boundaries that define whether a particular sea territory is under the control of a particular social group.

The fluidity of communal marine tenure is also apparent when considering the exclusivity of a territory. Although the exclusivity of the sea territory is formally defined in terms of the level of exploitation and the distribution of the catch, in practice the application of these criteria is often subjective (Chapters Five and Eight). Therefore, it should not be surprising if we find that someone who only does non-commercial fishing—an activity which is defined as 'free for all'—is driven from a sea territory because he does not have a good relationship with the owners of the territory. By contrast, we might find that a fisherman who is clearly fishing for commercial purpose is allowed to do so in another's sea territory because he is a good friend of the village head or other members of the right-holding unit.

What explains this fluidity? For one thing, the practice of communal marine tenure—as well as communal land tenure—is based on narratives of origin (*toom*) that are open to multiple interpretations. Hence, it often happens that more than one narrative of origin exists describing a particular issue or territory. These circumstances create multiple interpretations of boundaries as well as who are the appropriate right-holding units. Another important explanation is that the practice of marine tenure is embedded in the social structure of the community which is dynamic and constantly changing. Thus, even when people talk about the traditional concepts of community, definitions and developments introduced by external forces such as the modern state are also taken into account. Conversely, when the locals talk about the community, although they may refer to modern structures such as the village, people still utilize the traditional structures and terms. This is because people consider the introduction of modern formal structures—whether they be Dutch, Indonesian, Muslim or Christian—to be additions or enrichments to rather than replacements of the old structures (Chapters Two, Six, and Nine). Therefore, there are no apparent boundaries between traditional and modern constructions. It is a continuum like the flow of a river.

The embeddedness of communal marine tenure in the social structure of the community is apparent in contestations between social groups within the community. Contestation between and within different social ranks—which is seemingly one of the main ways by which the community keeps changing—usually concern control over the political and territorial domains within the village. Traditionally, control over each domain was separate with control over the territorial domain under the lord of the land who was a free person, and politics under the control of the traditional village head who was a noble. However, modern village organisation does not differentiate between these two domains. In the modern conception, those who control village politics are assumed to be the leaders of all matters related to the political and territorial domains of the village. Conflict over coastal boundaries between Sather and Tutrean and the issue of who has the rights to lease sea territory to an outside fishing company in Dullah Laut are examples of how the practice of marine tenure becomes an integral part of the contestation between free and noble villagers as well as between various groups within the nobility (Chapters Six and Nine). These conflicts are also examples of the inseparability of marine tenure and village politics.

Further, the embeddedness of marine tenure in the social structure of the community leads me to conclude that people do not always consider marine tenure when considering the relationship between people and the marine environment. In particular circumstances, control over marine territory is used in situations of social contestation and is considered to be an index of precedence. In this sense, people do not consider the practices of marine tenure to have anything to do with sustainability or the fair distribution of marine resources.

In addition, considering the role of marine tenure and its implementation in the discourse on community-based management and local ecological knowledge (LEK) and practices, it really problematises the validity of the discourse's basic premises. The assumptions 'that (1) local populations have a greater interest in the sustainable use of resources than does [sic] the state or distant corporate managers, that (2) communities are more congnisant of the intricacies of local ecological processes and practices, that (3) communities are more able to effectively manage those resources through local or traditional forms of access' (Tsing, Brosius and Zerner 2005: 1) and that (4) one of the key attributes of LEK is a people's shared system of knowledge or other expression about the environment and ecosystem relationships (Davis and Ruddle 2010) are questionable. In fact, discourse on the issues revealed that people used marine tenure more for political and social reasons at the expense of resource sustainability or a just resource distribution. These findings question the validity of the first three premises. Because marine tenure is also subject to contestation and disputes indicate that the practice not only means different things to different people, but that the meaning is also contradictory between different people. This creates inconsistencies within the fourth premise. Without reconsidering these issues, the problems are carried over to the more advanced discourse on collaborative management that will now be the focus of my discussion.

Communal Marine Tenure: Co-Management Problems

As noted in the introductory chapter, co-management is currently the most popular discourse of marine tenure as well as other common pool resources management. In theoretical terms, co-management is the result of a review of highly centralised and community-based resource management and its supporting theories. In practical terms, co-management refers to the management practice of government and fishing communities working together in crafting, implementing, and evaluating policy related to marine tenure. This means that there should be some transfer of power and obligation between government and the community. In this regard, government acknowledgement or even legalisation of the community management and rights system has been proposed as a way of creating co-management practices.

The lessons from the practice of communal marine tenure in the Kei Islands suggest that there would be problems in applying this principle. For example, if we start with the government's recognition of traditional marine tenure, I am afraid that government acknowledgement would generate further conflict. This is because as the case studies demonstrate, marine tenure is a contested practice. In such circumstances, government acknowledgement might only stimulate conflicting economic and political claims among segments of the community over the sea territory. While the economic interests relate to the benefits to be derived from the extraction of the resources, the political interests related to an acknowledgment means that the state recognises the special association between a particular segment of the community and the marine resource. Of course, this could become excellent ammunition for members of the community that are in contestation with others. Thus, the empowerment of tradition through formal government acknowledgement might lead to the risk of abuse. As previously noted, both traditional and 'modern' (government) leaders have been known to use tradition for their own benefits. This is what is popularly known as the 'elite capture phenomena' (Béné et al. 2009).

Furthermore, local conflicts not only demonstrate that the community is not homogenous or 'a unified, organic whole' (Agrawal and Gibson 1999) but that they are 'riven with differences in status' (Allison and Ellis 2001). The conflicts also show the presence of conflicting interests among the various segments of the community. This raises the following question: to whom in the community should management responsibilities be delegated? In conflicts over precedence within the community, people are often more concerned with gaining and holding power than respecting others in an equal relationship. In fact, the conflicts reveal the tendency of certain groups within the community to claim their superior status over others. Further, even if we assume that we can choose a representative from every segment of the community, negotiating sustainable and socially-just resource management remains problematic because segments within communities might create alliances with external agencies. (cf. Agrawal and Gibson 1999). Such alliances are not always in accordance with the benefit of the whole community or sustainable resource management. The case studies in this book show the tendency of particular segments of the community to trade their rights over the sea for political and economic support from local bureaucrats and the fishing industry. The key problem in developing systems of representation is how to avoid the collusion of some stakeholders to achieve their goals at the expense of other stakeholders. This book shows that issues of representation and community bargaining power cannot be separated from social differentiation, conflict within the community, and connections with external agencies.

Communal Marine Tenure and the Current Problems of Indonesian Fisheries

The discourse on traditional marine resource management in Indonesia has basically consisted of criticism of the failure of centralised government to promote sustainable use and fair distribution of resources. 'Government' here refers to the New Order Regime led by former president Soeharto who accumulated political power to control people and resources. New laws, regulations, and institutional arrangements were created to replace the supposedly deleterious structures of Soeharto's New Order Regime after it ended in 1998. It was then declared that Indonesia had entered an era of reform.

The maritime and fisheries sectors were among those subject to these changes and marine tenure is one of the issues that has been reformed. The *Law of Local Government* or *The Law of Local Autonomy*[1] (Law No. 22, 1999) has delegated power to the local govement to manage up to 12 miles of marine territories and resources. The authority to manage the first third of these areas is held by the district government or municipality, and the rest are managed by the provincial government. These articles introduce a new practice of marine tenure in Indonesia and draw boundaries between three different parts of the sea: (1) the sea territory from the coastal line to four miles;[2] (2) the sea territory from four

---

1 This law was revised by the Law No. 32, 2004. However the new law still maintains the articles referring to the distribution of management rights over the sea.
2 This is based on the assumption that the provincial sea territory is 12 miles, which is not always the case because some provinces are separated by less than 24 miles. For these provinces, the sea territory is less than 12 miles which means that because the law defines the district's sea territory to be one third of the province's territory, the district sea territory would be less than four miles.

to 12 miles; and (3) the sea territory from 12 miles to the Exclusive Economic Zone (EEZ).[3] The articles also define the right-holding units for each segment of the territory.

There has been a mixed response to this new marine tenure policy. The response has created conflict between fishermen who used to fish outside of their administrative sea territory (see for example Anonymous 2000a, b, d, e, f, and g). A well-reported conflict occurred between fishermen from the northern coast of Java and fishermen from Masalembo on the island of Madura (Anonymous 2000h). Driven by their belief that the fishermen from northern Java had encroached upon their sea territory, the fishermen from Masalembo captured and burned a Javanese boat. Fishermen from the northern coast of Java swept their sea territory in order to find Masalembo fishermen with the intention of burning one of their boats in return. This conflict and similar conflicts in other part of Indonesia[4] suggest that the main issue was the economic interest in excluding others from exploiting the same resources. None of the reports mention that ecological or environmental concerns were a driving factor for this exclusion. Although for a short period of time, the conflict decreased the pressure on the resources—since those who were driven away could not come back immediately—I do not think that the local community was automatically stimulated to consider the sustainability of the resource.

Interestingly, the reasons given by the Madurese fishermen when they raided the Javanese fishermen operating in Madurese waters were from both the *Law of Local Autonomy* and traditional marine tenure practices. Again, as in the Kei islands (Chapters Two, Six, and Nine) people seemingly considered the introduction of a new system by the government as an addition to the existing traditional practice. Such circumstances add to the complexity of marine resource management problems.

The implementation of the *Law on Coastal and Small Islands Management No. 27/2007* might also exacerbate this problem primarily because there is inconsistency in the law itself. Article 16 (1) notes that the use of the marine coastal area is given in the form of Use Right of Marine Coastal Area (*Hak Pengusahaan Perairan Pesisir* or HP3). This right covers the use of a marine coastal area from the surface up, and down to the bottom of the sea [art. 16 (2)]. Article 18 notes that government can grant the HP3 to any Indonesian citizen, any company (established according to the Indonesian law) or traditional society.[5]

---

3   The outer boundary is not mentioned in the law because the law only pertains to the distribution of the sea territory within Indonesian administrative sea borders.

4   Some examples were the conflicts between fishermen from Gresik and Mojokuto in Jawa and fishermen from Jakarta, and between fishermen from the northern coast of Java and those from Madura as highlighted in Adhuri (2003 and 2009) and Fox et al. (2005).

5   In February 2010, a coalition of NGOs, community organisations, and indivual fishermen lodged a legal request to Mahkamah Konstitusi (the Constitutional Court) to cancel these articles. After more than a year of

These articles obviously assume that the government controls marine coastal areas and are an attempt to stimulate competition between the three groups to aquire HP3 from the government. This clearly means that communal marine tenure is overlooked. On the other hand, article 61 (1) stipulates that government acknowledge, respect, and protect traditional communities', traditional fishing communities' rights, and local wisdoms over coastal[6] and small islands areas. Contrary to other articles on HP3, this article clearly provides the basis for coastal communities to exercise their customary marine tenure and associated management practices. Secondly, even if HP3 only covers coastal marine areas that are not under the traditional claim as argued by some of the leaders in the Ministry of Marine Affairs and Fisheries, a blind adoption of communal marine tenure could be problematic. The success of the revitalisation of traditional or 'pre-existing management systems' (Ruddle and Satria 2010) in other parts of Indonesia is attributed to the re-contextualising of the system to the specific needs and circumtances of the fisheries sectors (Satria and Matsuda 2004; Adhuri 2009; Satria and Adhuri 2010). This means that although the traditional system provides a sound foundation, some adjustments and reconfiguring are needed to convert it to a better resource management system.

Recent Developments in the Kei Islands

The collapse of the New Order Regime in 1998 brought about many changes including violent conflicts in 1999 and changes related to decentralisation policy. The following is an account of the roles of both local elites and tradition in these changes. At the end, I will reflect on ways in which these changes could affect the direction of marine resource management in the islands.

Violent conflict with religious underpinnings broke out in the Kei Islands in 1999. Unlike in Central and North Maluku where conflict lasted for years and was widespread, the conflict in the Kei Islands was short lived (three months) and concentrated in several places. The effectiveness of the reconciliation process taken during the conflict in the Kei Islands was attributed to the role of local tradition and traditional elites. For example, the conclusion of a minor conflict in the territory of Maur Ohoiwut—a kingdom on the northern part of Kei Besar Island—was due to strong leadership from the king and other traditional leaders in the domain (Laksono 2004). It was recounted that in order to avoid the conflict spreading to his territory, the king called most of the traditional leaders in his domain and explained the agreements, laws, and regulations that bind them together in harmony. He also insisted that they adhere to these traditions and

---

legal proceedings, the Contitutional Court agreed and cancelled them on 16 June 2011.
6   Coastal territory covers land (the land boundary of the coastal subdistrict) and marine (up to 12 miles) areas.

that as a result, no one should be involved in the conflict. The traditional leaders brought these messages back to their people and some of them arranged a joint patrol in their territory which saved their communities from any bloodshed.

The same strategy was also developed and implemented for reconciliation processes to heal the rifts between villagers after the conflicts in the Kei Islands. This time led by local elites, traditional songs were song, narratives of origin were chanted, and old rituals were performed. Indeed, this was the time when tradition was re-discovered and revitalised (Thornburn 2005). These traditions re-emphasised the importance of 'being self-corrective and acknowledging the presence of rightness in others', reminding the law of Hukum Larvul Ngabal and refreshing ancestors' agreements of peace and reconciliation (Elmas 2004; Laksono 2004; Kaplale 2004; Ngamelubun 2004 and Silubun 2004). This strategy was considered to be one of the main elements that brought peace back to the people of Kei Islands.

The fall of Soeharto's New Order Regime was followed by a policy of decentralisation in Indonesian politics. The central government, facilitated by the implementation of the *Law of Local Government* No. 22/1999 in 2000, transfered the authority to govern territory and people to local (particularly district and municipal) governments. The revision of Law No. 22/1999 and its replacement with Law No. 32/2004 has switched the election of district and municipal heads (as well as governors at provincial levels) from local parliament members to the people. Both changes have opened the door for local elites to play a more significant role in local political dynamics.

Making use of the decentralisation, Kei Islands elites succeded in changing the Kei Islands from two subdistricts—Kei Kecil and Kei Besar—into one district and one municipality, Maluku Tenggara District and Tual Municipality respectively. This means that with the consent of the central government, they created a new government body—the Tual Municipality. They also kept the Maluku Tenggara District from covering the Aru Islands and Maluku Tenggara Jauh leaving it to exclusively cover the Kei Islands.[7] At the subdistrict levels, the Kei Kecil subdistrict was split into seven subdistricts and Kei Besar became three subdistricts (ICG, 2007).

Interestingly, tradition also plays an important role and was instrumental in local political contestations. For example, in the fight over the seat of head of Maluku Tenggara District that took place in 2003, it was noted:

> On 16 September, invoking *adat*, the *raja* of Tual, from the same family as the losing Golkar candidate, M.M. Tamher, ordered that fences of

---

7   Maluku Tenggara Barat and Aru Islands subdistricts split from Maluku Tenggara District and upgraded to become independent districts in 2000 and 2003 respectively.

coconut leaf (*hawear*) be placed around the *bupati*'s office and elsewhere in Tual, including the airport, harbour and the major bridge linking the islands of Kei Kecil and Dullah, disrupting schools, commerce and transport. Sometimes defined as "*adat* no trespass signs", the fences are considered impassable under *adat* law: those who dare to cross them risk "the wrath of unknown forces" (ICG 2007: 6).

This protest was directed at the winning candidate who was accused of manipulating the election using the support of President Megawati's husband and a Jakarta businessman who owns the largest fisheries company—PT. TJ—on Dullah Island. Although, this protest did not affect the appointment of the winning candidate, it created a very tense atmosphere in the district.

To give another example, in 2009 22 traditional kings in the Kei Islands representing the five and nine groups awarded the director of PT. TJ a traditional title called 'who stands in front and distributes' or *primus inter pares* (*dir u ham wang*). (Anonymous 2000c, Hooe 2012). Traditionally, this title was granted to a distinguished leader who would make a positive contribution to the order and lives of many in the Kei Islands. The holder of this title is also considered as powerful as any of the leaders. The appointment was performed with a traditional ceremony in the largest field in the centre of the district, which brought about a huge protest from various elements of the community. Thousands of people from various backgrounds, led by the Muslim University Association, blocked the main road in Tual. The protest was not only driven by anger against the traditional leaders for abusing tradition, but also for possible abuses by the director of PT. TJ. There were rumours that the traditional leaders had been bribed by the director of PT. TJ., and that this award was considered the equivalent of giving PT. TJ the monopoly on importing staple foods such as rice, sugar and coffee to the Kei Islands. Above all, since PT. TJ was considered a large fisheries company, people feared that giving this title to its director would lead the company to essentially control the marine territory and fisheries. It was the fear of a recurrence of the violent conflicts of 1999 that forced the district head to cancel the appointment.

What affect have these changes brought about by the end of the New Order Regime had on the practice of marine resource management in the Kei Islands? The short term consequences of the 1999 conflict were that all of the outside-owned fishing companies were driven away and a freeze was placed on coral reef fish and anchovy export. The presence of outside fishing companies and pressure from the international market had been leading to higher levels of exploitation and conflict in the islands. When these companies left and the freeze put into effect, the absence of external fishing companies, the disruption of the international fish trade, and the absence of fisheries conflict coupled with the successful reconciliation of the 1999 conflict provided an opportunity for locals to reconsider and re-craft a better approach to marine resource management. The government's decentralisation policy also provided a chance for villagers in the Kei Islands to better manage their resources. When reflecting on the ways in which locals used tradition to avoid conflict and for reconciliation, it's apparent that local elites and villagers have proven their ability to move in the direction where common goals and interests override self interests, and where tradition can become an effective tool. However, we also see from the political games of the post-New Order Regime that local elites do not stop using tradition for their own gain and at the cost of public interest. This means that constestation for power and control over resources is still rampant, and that tradition can be utilised for both good and selfish causes. The future of resource management in Kei Islands will depend on which of the two uses is more frequent.

Appendix 1

This appendix details an economic analysis of traditional fishing practices in Dullah Laut. Using examples from fishing practices in use during my field research, the discussion will include investment and return details of each type of traditional fishing gear employed.

Fish pots

The investment for fish pot fishing—or *vuv*—is relatively small, about Rp4000 per pot. Fishermen buy parts for the fish pot for as little as Rp3000 in Dian Island and need an additional Rp1000 for transportation. Fish pot fishermen usually have 15 to 25 fish pots operating at any one time. This requires an investment of around Rp60 000 to Rp100 000. Those who worked in groups operated up to 100 fish pots at a time. An example of the latter was a group formed by three persons who used 60 fish pots requiring an investment or Rp240 000.

I noted at least three ways fishermen collected the money to buy fish pots. The first was from their business income which might include income from other fishing activities or from selling agricultural products. The second was from borrowing money from a village middleman to whom they sold their catch. Loan repayments were usually deducted from the price of their catch and this kind of arrangement was not strict in terms of the level of repayments or the period of the debt. The third way was to form a group in which one of the members provided the money to buy the fish pots while others contributed their labour to operate the business.

Fishermen believed that the return from fish pot fishing was relatively small but that it was consistent with the low level of investment. The operation of fish pots was also considered to be low risk. The following two cases illustrate these statements.

Case 1:[1] Mr Yaum operated 15 fish pots from March until May 1996. In March he went fishing for 26 days and earned Rp268 500. In the second month—April— he only went fishing for twenty days and earned Rp198 500. May was the last month he used the same fish pots because at the end of this month most of them

---

1   It was difficult to get complete catch data from a fisherman—or a group of them—over a long period. Most if not all fishermen who worked alone did not record their catch. Those who worked in groups only kept monthly notes and once the catch was distributed to the group, the note would be thrown away. The data used in this section was based on a form that I left with the fishermen for them to fill in. It was an empty table

were damaged and unrepairable. In this month, he went fishing for 24 days and earned Rp205 500. Based on this, we can calculate that from three months use of his 15 fish pots, Mr Yaum's total income was about Rp672 500 (Table A-1). Deducting Rp60 000 for buying and transporting the fish pots, his net income for three months from 15 fish pots was Rp612 500.

Case 2: Daeng, Asis, and Saleh formed a fishing group. Daeng provided the group with Rp161 000 for 46 fish pots. Asis and Saleh operated the fish pots. They operated all of the *vuv* for one month and achieved a catch worth Rp430 000. They stopped fishing in the second month due to fasting. When they went to use their fish pots after Ramadhan, twenty of the fish pots were rotten[2] and the rest could only be used for two weeks. However, they still earned Rp150 000. This meant that their total catch was Rp580 000. Based on the agreement that the three members were considered to have contributed equal shares, the total return was divided into three totalling Rp193 000 each. For both Asis and Saleh, this was their net income while for Daeng the net income was only Rp32 000 as he had invested Rp161 000 for the fish pots (Table A-1).

Table A-1: The economy of fish pot fishing.

	Investment (Rp)	Return (Rp)
Case 1		
15 fish pots	60 000	
March		268 500
April		198 500
May		205 500
Total	60 000	672 500
Net return		612 500
Net return per month		204 166
Case 2		
46 fish pots (Daeng)	161 000	
First month's catch		430 000
Two weeks catch		150 000
Total		580 000
One third share		193 000
Net return for Daeng		32 000
Net return for Asis and Saleh		193 000

Source: Fieldwork research.

noting the running costs and the return for every time he/they went fishing for a month. Unfortunately not all fishermen filled in the form regularly. However, the available data was enough to get a reasonable idea of the economy of fishing in Dullah Laut Village.

2   This was common because the fish pots were made of young bamboo which was easily damaged particularly when on land, I was told.

Stake traps

A stake trap is usually locally constructed. With the help of two or three people, a stake trap takes three or four days to prepare. A fisherman usually buys the bamboo from another village on Dullah or Kei Kecil Island then he cuts, smooths and plaits the bamboo in Dullah Laut Village. The plaiting is usually done with a rope made from vines available in the bush at Dullah Laut. Wooden stakes are also collected from bush or forest. They are needed to provide the frame of the stake trap to which the plaited bamboo is attached. Once this preparation is complete, fishermen bring the bamboo and stakes to the location where the trap is to be set up. Tree stakes are planted first then the bamboo fence is tied to them. A fisherman checks his stake trap every day and collects the catch if there is any. The trap lasts for about six months before the bamboo deteriorates.

There were not many stake traps operating during the period of my fieldwork. I observed only seven and four stake traps being used during the west and east monsoons respectively. When I asked people whether this this was the norm, it was explained that this had been the case for the last two decades. According to an informant, if I had done my fieldwork in the 1970s I would have found more than 15 stake traps operating in Dullah Laut. Now, he explained, fish had become 'clever' and avoided the stake trap.[3] For the last two decades the stake trap did not catch many fish and the investment needed to construct a stake trap had increased significantly. Accordingly, some fishermen gave up stake trap fishing. A young fisherman told me that although line fishing required more energy, it was much better than stake trap fishing as the investment was not as much for stake traps. Also, he explained that stake trap fishing was static—waiting for the fish to enter the trap. Line fishing on the other hand, was dynamic. He could choose any fishing spot and move to another if the fish were not biting. In justifying why stake trap fishing was best, an older fisherman explained he only needed three or four days of hard work to construct the stake trap which provided six months of minimal effort harvesting. Regarding the amount of fish that were caught, he commented that it was God's will. Comparing the efforts, the stake trap fisherman spent an hour to collect the catch while line fishing required at least three or four hours to get a catch of similar size.

The economic aspects of stake trap fishing are given in Table A-2. These calculations demonstrate that the investment needed for stake trap fishing is much larger than for fish pot fishing. At least Rp250 000 is needed for the bamboo plus food, drink, and cigarettes for those helping to construct the stake trap. Although I was not able to collect economic data of a particular

---

3   When I asked whether this was an indication of over fishing, my informant told me that fish were still abundant in the location but were not getting trapped.

stake trap for a six-month period—which is the usual life of a stake trap—I did collect some data on the catches of five fishermen (for a one-month period) over five months. The data in Table A-1 provides a general picture of the economic returns of stake trap fishing indicating the average monthly return from a stake trap was higher than the fish pot.

Table A-2: The economy of stake trap fishing.

	Investment (Rp)	Return (Rp)
Bamboo	200 000	
Food, drink, cigarettes	50 000	
July 1996		330 000
August 1996		563 500
September 1996		304 250
January 1997		197 750
February 1997		133 500
The sixth month*		305 800
Total	250 000	1 834 800
Net return		1 584 800
Net return per month (over 6 months)		264 133

* The 'catch' for this month is the average of the other five months.

Line fishing

To shed some light on the economy of line fishing, two examples are summarised in Table A-3.

Case 3: Mr Ato mostly employed bottom line fishing and troll line fishing[4] during the period of my fieldwork. His income from line fishing for three consecutive months (November 1996 to January 1997) was as follows. In November 1996 when he went fishing for 22 nights (four nights troll line fishing and 18 nights bottom line fishing), he earned Rp441 750. In December 1997 when he went fishing for 22 nights (three nights troll line fishing and 19 nights bottom line fishing), his income was Rp281 000. His income on the third month, January 1997, when he went bottom line fishing almost every night and only used a troll line once, was Rp85 000.[5] This meant that for those three months his total income was Rp807 750. According to Mr Ato, he spent about Rp5000 on sinkers

---

4 This technique involves towing the line behind the canoe.
5 According to Mr Ato, although he went fishing every night, he did not spend as many hours as he had during other months because Ramadhan started on 11 January.

and hooks that he lost during fishing. He did not buy any line during these three months, nor did he buy fuel since he used a small canoe which he paddled. This meant that his net average income was Rp267 583 per month.

Case 4: In June 1996, Mr Abd went fishing every night except for one. His total catch came to Rp586 000. In July 1999, Abd went fishing with Ish using his canoe which was equipped with an outboard engine. Over 21 nights fishing, they caught Rp655 000 worth of fish. After deducting the price of fuel used for the engine and sharing the catch equally with Ish, Abd's net income was Rp188 000. In August 1996, Abd, who did not go fishing together with Ish any more, went fishing for only 9 nights due to bad weather. His catch was sold for Rp87 500. His total income for the three months was Rp861 500 but was not the net value of his fishing because according to Abd, he spent around Rp10 000 on fishing gear. This meant that the net value was Rp851 500, an average return of Rp283 833 per month.

Table A-3: The economy of line fishing.

	Investment (Rp)	Return (Rp)
Case 3		
Sinkers and hooks	5000	
November 1996		441 750
December 1996		281 000
January 1997		85 000
Total	5000	807 750
Net return		802 750
Net return per month		267 583
Case 4		
Fishing gear	10 000	586 000
June 1996		188 000
July 1996		87 500
August 1996		
Total	10 000	861 500
Net return		851 500
Net return per month (over 3 months)		283 833

Source: Fieldwork research.

Net fishing

In detailing the economic analysis of net fishing, I will highlight the cases of two groups of fishermen from Dullah Laut Village.

Case 5: Daud's group has five fishermen who had been involved in net fishing since 1983. Mr Daud, who was also the owner of the fishing gear, led the group. The crew was made up of two of his brothers and two of his cousins. Mr Daud's investment in the fishing gear between 1982 and 1997 was Rp8 410 000 (see Table A-4). During my fieldwork this fishing gear was still in use. According to Mr Daud's calculations, the fishing gear would last 15 years even though there was already significant deterioration evident. For example, the original net he had bought in 1982 was almost completely destroyed. Although the net was mostly made up of new netting, Mr Daud still considered it to be the net he bought in 1982. I also observed that his outboard engine had undergone numerous repairs. Nevertheless, I believed Mr Daud's investment calculation was still reasonable, given repairs to the net and engine are calculated separately. On this basis the average capital needed to operate his net fishing activities was Rp560 666 per year.

Table A-4: Daud group net fishers: Capital requirements.

Fishing gear	Quantity	Purchase date	Cost (Rp)
Boat	1	1982	150 000
Net	5	1983	250 000
8HP Outboard engine	1	1986	2 600 000
Net	4	1992	300 000
Boat	1	1992	300 000
Net	1	1993	200 000
Net	3	1994	360 000
Net	10	1995	250 000
15 HP Outboard engine	1	1996	4 000 000
Total capital			8 410 000
Capital per year (over 15 years)			560 666
Capital per month (over 12 months)			46 722

Source: Fieldwork research.

The running costs and the catch value for the period November 1996 to May 1997 are set out in Table A-5.

Table A-5: Daud group net fishers: Running costs and fish catch returns.

	Running costs (Rp)		Catch value (Rp)
Fuel	4 200 000	Nov 1996	3 536 850
Net repair	700 000	Dec 1996	2 052 700
Engine maintenance	1 050 000	Jan 1997	2 963 450
Boat maintenance	100 000	Feb 1997	1 972 450
		Mar 1997	1 355 250
		Apr 1997	2 994 300
		May 1997	3 307 100
Total	6 050 000		18 182 100
Amount for distribution (minus fuel costs)			13 982 100
Owner's share (40%)			5 592 840
Crew's share (60%)			8 389 260
Individual (7) crew share			1 198 465
Monthly individual crew share			171 209

Source: Fieldwork research.

The economic return differed quite markedly for the owner of the gear and crew. For the crew, their income was calculated by means of simple distribution. After deducting the cost of the fuel, the value of the catch was distributed into two shares with the owner receiving a 40 per cent share, and the crew receiving a 60 per cent share which was distributed to each member of the group (including the crew leader, also owner of the gear) equally. In this case, the owner's share was Rp5 592 840 and the income of each crew member was Rp1 198 465. The average monthly income for the owner was Rp798 977 and each individual crew member had a net monthly income of Rp171 209. The calculations used to determine the net income of the owner have been provided in Table A-6.

Table A-6: Daud group net fishers: Net owner's income.

	Investment (Rp)	Return (Rp)
Capital for 7 months (Table A-4)	327 054	
Non-fuel running costs (Table A-5)	1 850 000	
Owner's share		5 592 840
Crew member's share		1 198 465
Total	2 177 054	6 791 305
Net return for 7 months		4 614 251
Net return per month		659 178

Source: Fieldwork research.

Case 6: In 1985, Mr Saban was given a 40-metre long net by a relative from another village. This gift prompted Mr Saban to buy gear needed to operate the net. In the same year he bought a boat, an outboard engine, and five additional nets each 8 m long. When his outfit was ready, Mr Saban asked his son and two of his nephews to join him. They started their net fishing in 1985 and were still fishing when I finished my fieldwork in 1997. According to Mr Saban, he anticipated using the gear for another three years—a total life of about 15 years. His total capital expenditure for the net operation was Rp2 250 000 which represented a return of Rp12 500 per month.

Table A-7: Saban group net fishers: Expenditure and returns.

	Quantity	Cost
Boat	1	450 000
Net	5	300 000
Outboard engine	1	1 500 000
Total capital		2 250 000
Capital per year (over 15 years)		150 000
Capital per month		12 500

Source: Fieldwork research.

I was able to record Mr Saban's group fish catches and running costs continuously over a period of nine months (April to December 1996). These records are summarised in Table A-8.

Table A-8: Saban group net fishers: Costs and catch records.

	Running costs (Rp)	1985	Catch value (Rp)
Fuel	2 265 000	April	622 300
Net repair	450 000	May	816 000
Engine maintenance	950 000	June	890 700
Boat maintenance	100 000	July	1 232 150
		August	1 429 000
		September	208 000
		October	2 205 800
		November	2 583 100
		December	619 000
Total*	1 500 000		10 606 050
Amount for distribution (minus fuel costs)			8 341 050
Owner's share (40%)			3 336 420
Crew's share (60%)			5 004 630
Individual crew (4) share			1 251 158
Monthly individual crew share			139 018

Source: Fieldwork research.

* For non-fuel running costs. In calculating Mr Saban's net return, taking into consideration his capital and running costs, his net income totals Rp330 564 per month (see Table A-9).

Table A-9: Saban group net fishers: Net owner's income.

	Investment (Rp)	Return (Rp)
Capital for 9 months (Table A-7)	112 500	
Non-fuel running costs (Table A-8)	1 500 000	
Owner's share		3 336 420
Each crew member's share		1 251 158
Total	1 612 500	4 587 578
Net return for 9 months		2 975 078
Net return per month		330 564

Source: Fieldwork research.

Bibliography

Adhuri, D.S., 1993. 'Hak Ulayat Laut dan Dinamika Masyarakat Nelayan di Indonesia Bagian Timur: Studi Kasus di P. Bebalang, Desa Sathean dan Demta [Communal Marine Tenure and the Dynamics of Fishing Societies in Eastern Indonesia: Case Studies in Bebalang Island, Sathean and Demta Villages].' *Masyarakat Indonesia* 20(1): 143–163.

———, 1998a. 'Who Can Challenge Them? Lessons Learned from Attempting to Curb Cyanide Fishing in Maluku Indonesia.' *Live Reef Fish Information Bulletin* 4: 12–17.

———, 1998b. 'Saat Sebuah Desa Dibakar Menjadi Abu: Hak Ulayat Laut dan Konflik Antar Kelompok di Pulau Kei Besar [When a Village was Burnt to Ashes: Communal Marine Tenure and Group Conflict in Kei Besar Island].' *Antropologi Indonesia* 57: 92–109.

———, 1999. 'The Incident in West Monsoon: Marine Tenure and the Politics of *Kepala Desa*.' Paper read at the 5th Maluku Conference, Darwin, 14–16 July.

———, 2001. 'Antara Ikan Garopa dan Otonomi Daerah: Politik Manajemen Sumberdaya Laut [Between Grouper Fish and Local Autonomy: The Politics of Marine Resource Management].' *Antropologi Indonesia* 65: 84–95.

———, 2003. 'Does the Sea Divide or Unite Indonesians? Ethnicity and Regionalism from a Maritime Perspective.' Canberra: Research School of Pacific and Asian Studies, Resource Management in Asia-Pacific Program (Working Paper 48).

———, 2004. 'The Incident in Dullah Laut: Marine Tenure and the Politics of Village Leadership in Maluku, Eastern Indonesia.' Maritime Studies (MAST) 3(1): 5–23.

———, 2005. 'Perang-perang atas Laut, Menghitung Tantangan pada Manajemen Sumberdaya Laut di Era Otonomi: Pelajaran dari Kepulauan Kei, Maluku Tenggara [Wars on the Sea: Calculating the Challenges for Decentralized Marine Resource Management: Lessons from Kei Archipelago, southeastern Maluku].' *Antropologi Indonesia* 29: 300–308.

———, 2009. 'Social Identity and Access to Natural Resources: Ethnicity and Regionalism from a Maritime Perspective.' In M. Sakai, G. Banks and J.H. Walker (eds), *The Politics of the Periphery in Indonesia: Social and Geographical Perspectives*. Singapore: National University of Singapore Press.

Agrawal, A. and C.C. Gibson, 1999. 'Enchantment and Disenchantment: The Role of Community in Natural Resource Conservation.' *World Development* 27: 629–649.

Allison, E.H. and F. Ellis, 2001. 'The Livelihoods Approach and Management of Small-Scale Fisheries.' *Marine Policy* 25: 377–388.

Anon., 1991. 'Laporan Penelitian Hak Adat Kelautan di Maluku [Report of Research on Customary Marine Rights in Maluku].' Ambon: Pattimura University and Hualopo Foundation.

———, 2000a. 'Nelayan Madura Mengusir Nelayan Pantura [Madurese Fishermen Drove Away Fishermen from Pantura].' *Kompas*, 25 February.

———, 2000b. 'Hari ini Nelayan Pantura akan Demo ke Jakarta [Pantura Fishermen Will Demonstrate to Jakarta Today].' *Republika*, 13 November.

———, 2000c. 'Nelayan Pantura Jawa Protes Laut "Dikapling" [Pantura Fishermen Protest over Marine "Plotting"].' *Kompas*, 14 November.

———, 2000d. 'Nelayan Desak Pemerintah Keluarkan Perpu Untuk Tangkap Ikan.' ['Fishermen Force Government to Issue the Law for Catching Fish.'] *Suara Pembaruan*, 14 November.

———, 2000e. 'Peta Wilayah Laut Tidak Untuk Membatasi Nelayan.' ['Marine Map is not for Limiting Fishermen.'] *Suara Pembaruan*, 16 November.

———, 2000f. 'Gawat, Laut Indonesia Sudah di Kavling [Indonesian Sea has been Plotted].' *Rakyat Merdeka*, 17 November.

———, 2000g. 'Pembakaran Itu... [That Burning...].' *Rakyat Merdeka*, 17 November.

———, 2000h. 'Pemerintah Nggak Siap [The Government was not Ready].' *Rakyat Merdeka*, 17 November.

———, 2009. 'Ribuan Warga Tual Protes Gelar Adat: Warga Mencurigai Monopoli Bisnis Perikanan [Thousands of People Protest the Award of Traditional Leadership Title: People Suspected Fisheries Business Monopoly].' *Kompas*, 18 October.

———, n.d. 'Kumpulan Perundang-Undangan/Peraturan Rerikanan Laut [The Compilation of Fisheries Laws/Regulations].' Yasamina.

Antunès, I., 2000. Le Développement Local de la Pêche en Indonésie : Entre Unité Politique et Diversité Culturelle [Local Fisheries Development in Indonesia: Between Political Unity and Cultural Diversity]. Paris: Université de Paris IV Sorbonne; Sydney: Sydney University (Ph.D. thesis).

————— and S.A.P. Dwiono, 1998. *Watlar, an Eastern-Indonesian Village Caught Between Tradition and Modernity.* Montpellier: Centre ORSTOM.

BPSK (Badan Pusat Statistik Kabupaten), 2000. 'Maluku Tenggara Dalam Angka [Maluku Tenggara in Figures].' Tual: BPSK [District Statistical Centre].

—————, 2003. 'Maluku Tenggara Dalam Angka [Maluku Tenggara in Figures].' Tual: BPSK [District Statistical Centre]

Bailey, C., 1986. 'Government Protection of Traditional Resource Use Rights—the Case of Indonesian Fisheries.' In D.C. Korten (ed.), *Community Management: Asian Experience and Perspectives.* West Hartford (CT): Kumarian Press.

—————, 1988. 'The Political Economy of Marine Fisheries Development in Indonesia.' *Indonesia* 46: 25–38.

—————, 1997. 'Lesson from Indonesia's 1980 Trawler Ban.' *Marine Policy* 21: 225–235.

————— and C. Zerner, 1992. 'Community-Based Fisheries Management Institutions in Indonesia.' *Maritime Anthropological Studies* 5: 1–17.

Balland, J. and J. Platteau, 1996. *Halting Degradation of Natural Resources: Is there a Role for Rural Communities?* New York: Oxford University Press.

Barraud, C., 1979. *Tanebar-Evav: Une Société de Maisons Tournée vers le Large [Tanebar-Evav: A Society of Houses Facing the Ocean].* Cambridge: Cambridge University Press.

—————, 1990a. Kei Society and the Person: An Approach through Childbirth and Funerary Rituals. *Ethnos* 55: 215–231.

—————, 1990b. Wife-Givers as Ancestors and Ultimate Values in the Kei Islands. *Bijdragen Tot De Taal-, Land- En Volkenkunde* 146: 193–225.

Béné, C., E. Belal, M.O. Baba, S. Ovie, A. Raji, I. Malasha, F. Njaya, M.N. Andi, A. Russell and A. Neiland, 2009. 'Power Struggle, Dispute and Alliance over Local Resources: Analyzing "Democratic" Decentralization of Natural Resources through the Lenses of Africa Inland Fisheries.' *World Development* 37: 1935–1950.

Bavinck, M., 2001. *Marine Resource Management: Conflict and Regulation in the Fisheries of the Coromandel Coast*. New Delhi: Sage Publications.

Bedaux, C., 1978. *War Came to the Kai Islands*. Missionaries of the Sacred Heart.

Bellwood, P., 1996. 'Hierarchy, Founder Ideology and Austronesian Expansion.' In J.J. Fox and C. Sather (eds), *Origins, Ancestry and Alliance: Explorations in Austronesian Ethnography*. Canberra: Australian National University, Research School of Pacific and Asian Studies, Department of Anthropology.

Berhitu, 1987. 'Laporan Perkembangan Pelaksanaan U.U. No. 5 Tahun 1979 tentang Pemerintahan Desa [Report on the Progress of the Implementation of Law No. 5, Year 1979 on Village Government].' Kei Besar: Camat Kei Besar.

Berkes, F., (ed.), 1989. *Common Property Resources: Ecology and Community-Based Sustainable Development*. London: Belhaven Press.

Bezemer, T.J. (ed.), 1921. *Beknopte Encyclopaedie van Netherlandsch-Indie [Concise Encyclopaedia of the Netherlands Indies]*. Gravenhage: M. Nijhoff.

Brosius, J.P, A.L. Tsing and C. Zerner, 2005. 'Introduction: Raising Questions about Communities and Conservation.' In J.P. Brosius, A.L. Tsing and C. Zerner (eds), *Communities and Conservation: Histories and Politics of Community-Based Natural Resource Management*. Lanham (MD): Altamira Press.

Chauvel, R., 1985. *The Rising Sun in the Spice Islands: A History of Ambon during the Japanese Occupation*. Clayton: Monash University, Centre for Southeast Asian Studies.

————, 1999. 'Ambon's Second Tragedy: History, Ethnicity and Religion.' Paper read at the 5th Maluku conference, Darwin, 14–16 July.

Christy, F.T., 1982. 'Territorial Use Rights in Marine Fisheries: Definitions and Conditions.' Rome: Food and Agriculture Organisation (Fisheries Technical Paper 227).

Cooley, F.L., 1962. *Ambonese Adat: A General Description*. New Haven (CT): Yale University, Southeast Asian Studies Center (Cultural Report Series Volume 10)

————, 1973. 'Persentuhan Kebudayaan di Maluku Tangah [Cultural Contiguity in Central Maluku].' In P.R. Abdurrachman, R.Z. Leirissa and C.P.F. Luhulima (eds), *Bunga Rampai Sejarah Maluku[A Compilation of the History of Maluku]*. Jakarta: LIPI Centre for Scientific Documentation.

Crocombe, R. (ed.), 1974. 'An Approach to the Analysis of Land Tenure Systems.' In H.P. Lundsgaarde (ed.), *Land Tenure in Oceania*. Honolulu: University Press of Hawaii.

Crouch, H., 1979. Patrimonialism and Military Rule in Indonesia. *World Politics* 31: 571–587.

Davis, A. and K. Ruddle, 2010. 'Constructing Confidence: Rational Skepticism and Systematic Enquiry in Local Ecological Knowledge Research.' *Ecological Applications* 20: 880–94.

Dayton, L., 1995. 'The Killing Reefs.' *New Scientist* 148 (2003): 14–15.

Demsetz, H., 1967. 'Toward a Theory of Property Rights.' *American Economic Review* 57: 347–359.

DEZ (Departement van Economiche Zaken), 1936. 'Volkstelling 1930, Overzicht voor Nederlandisch-Indie [Summary of the 1930 Census in the Netherlands Indies].' Batavia: Landsdrukkerij.

Elmas, P., 2004. 'Perjalanan Menemukan Jati-diri: Menelusuri Jejak Konflik & Landasan Rekonsiliasi Dalam Masyarakat Kei [The Journey to Find Ourselves: Tracing the Conflict and Base for Reconciliation in Kei Society].' In P.M. Laksono and R. Topatimasang (eds), op. cit.

Feeny, D., F. Berkes, B.J. McCay and J.M. Acheson, 1990. 'The Tragedy of the Commons: Twenty-Two Years Later.' *Human Ecology* 18: 1–19.

Fox, J.J., 1988. 'Origin, Descent and Precedence in the Study of Austronesian Societies.' Public lecture presented at the University of Leiden, 17 March.

———, 1994. 'Reflections on "Hierarchy and Precedence"'. *History and Anthropology* 7: 87–108.

———, 1995a. 'Installing the "Outsider" Inside: The Exploration of an Epistemic Austronesian Cultural Theme and Its Social Significance.' Paper presented at the first European Association for Southeast Asian Studies conference, Leiden, 29 June–1 July.

———, 1995b. 'Origin Structures and Systems of Precedence in the Comparative Study of Austronesian Societies.' In P.J.K. Li, C. Tsang, Y. Huang, D. Ho and C. Tseng (eds), *Austronesian Studies Relating to Taiwan*. Taiwan: Academia Sinica, Institute of History and Philology.

————, 1996. 'Introduction.' In J.J. Fox and C. Sather (eds), *Origin, Ancestry and Alliance: Explorations in Austronesian Ethnography*. Canberra: Australian National University, Research School of Pacific and Asian Studies, Department of Anthropology.

————, 2008. 'Installing the "Outsider" Inside: The Exploration of an Epistemic Austronesian Cultural Theme and Its Social Significance.' Indonesia and the Malay World 36: 201–218.

————, D.S. Adhuri and I.A.P. Resosudarmo, 2005. 'Unfinished Edifice or Pandora's Box? Decentralization and Resource Management in Indonesia.' In B.P. Resosudarmo (ed.), *The Politics and Economics of Indonesia's Natural Resources*. Singapore: Institute of Southeast Asian Studies.

Geurtjens, H., 1911. 'De Slavernij van de Kei-eilanden [Slavery in the Kei Islands].' *De Java-Post*, 19 May.

————, 1921. *Uit Een Vreemde Wereld of Het Leven en Streven der Inlanders op de Kei-Eilanden [From a Strange World or The Life and Pursuing the natives on the Kei Islands]*. Hertogenbosch: Teulings.

Gordon, H.S., 1954. 'The Economic Theory of a Common Property Resource: The Fishery.' *Journal of Political Economy* 62: 124–142.

GBHD (Great Britain Hydrographic Department), 1943. *Eastern Archipelago Pilot – Volume III: Including the North-Eastern End of Celebes, Molucca, Ceram, Banda and Arafura Seas, and the Western End and Southern Coast of Netherlands New Guinea* (4th edition). London: HMSO.

GBNID (Great Britain Naval Intelligence Division), 1944. *Netherlands East Indies* (2 volumes). London: GBNID (Geographical Handbook Series).

Hardin, G., 1968. 'The Tragedy of the Commons.' *Science* 162: 1243–1248.

———— and J. Baden (eds), 1977. *Managing the Commons*. San Francisco: W.H. Freeman.

Hardjono, J., 1991. 'The Dimensions of Indonesia's Environmental Problems.' In J. Hardjono (ed.) *Indonesia: Resources, Ecology, and Environment*. New York: Oxford University Press.

Harkes, I. and I. Novaczek, 2002. 'Presence, Performance, and Institutional Resilience of *Sasi*, a Traditional Management Institution in Central Maluku, Indonesia.' *Ocean & Coastal Management* 45: 237–260.

Haverfield, R., 1999. '*Hak Ulayat* and the State: Land Reform in Indonesia.' In T. Lindsey (ed.) *Indonesia: Law and Society*. Leichardt (NSW): Federation Press.

Hooe, T.R., 2012. 'Little Kingdoms': *Adat* and Inequality in the Kei Islands, Eastern Indonesia. Pennsylvania: University of Pittsburgh (Ph.D. thesis).

Hviding, E., 1989. '"All Things in Our Sea": The Dynamics of Customary Marine Tenure, Marovo Lagoon, Solomon Islands.' Boroko (PNG): National Research Institute (Special Publication 13).

ICG (International Crisis Group), 2002. 'Indonesia: The Search for Peace in Maluku.' Jakarta: ICG (Asia Report 31).

―――, 2007. 'Indonesia: Decentralization and Local Power Struggles in Maluku.' Jakarta: ICG (Asia Briefing 64).

Jentoft, S., 1989. 'Fisheries Co-Management: Delegating Government Responsibility to Fishermen's Organizations.' *Marine Policy* 13:137–154.

―――, 2005. 'Fisheries Co-Management as Empowerment.' *Marine Policy* 29: 1–7.

―――, B.J. McCay and D.C. Wilson, 1998. 'Social Theory and Fisheries Co-Management.' *Marine Policy* 22: 423–436.

Johannes, R.E., 1978. 'Traditional Marine Conservation Methods in Oceania and Their Demise.' *Annual Review of Ecology and Systematics* 9: 249–364.

―――, 1981. *Words of the Lagoon: Fishing and Marine Lore in the Palau District of Micronesia*. Berkeley: University of California Press.

―――and M. Riepen, 1995. 'Environmental, Economic, and Social Implications of the Live Reef Fish Trade in Asia and the Western Pacific.' Jakarta: The Nature Conservancy.

Kaartinen, T., 2009a. 'Hierarchy and Precedence in Keise Origin Myths' In M.P. Vischer (ed.), *Precedence: Social Differentiation in the Austronesian World*. Canberra: ANU E Press.

―――, 2009b. 'Urban Diaspora and the Question of Community.' *Journal of the Finnish Anthropological Society* 34(3): 56–67.

―――, 2010. *Songs of Travel and Stories of Place: Poetics of Absence in an Eastern Indonesian Society*. Helsinki: Suomalainen Tiedeakatemia (FF Communications 299).

KSMT (Kantor Statistik Maluku Tenggara), 1993. 'Kecamatan Kei Kecil Dalam Angka [Kei Kecil Sub-District in Figures].' Tual: KSMT [Maluku Tenggara Statistical Office].

————, 1995. 'Maluku Tenggara Dalam Angka [Maluku Tenggara in Figures].' Tual: KSMT [Maluku Tenggara Statistical Office].

Kaplale, D., 2004. 'Ke Arah Rekonstruksi Etnis: Sejarah Pergulatan Politik Indentitas Kelompok di Maluku & Pelajaran dari Kei [Toward Ethnic Reconciliation: The History of Group Political Identity Struggle in Maluku & Lessons from Kei].' In P.M. Laksono and R. Topatimasang (eds), op. cit.

Kato, T., 1989. 'Different Field, Similar Locusts: *Adat* Communities and the Village Law of 1979 in Indonesia.' *Indonesia* 47: 89–114.

Kissya, E., 1995. *Sasi Aman Haru-Ukui: Traditional Management of Sustainable Natural Resources in Haruku*. Jakarta: Sejati Foundation.

Kristiadi, J., 1999. 'The Future Role of ABRI in Politics.' In G. Forrester (ed.), *Post-Soeharto Indonesia: Renewal or Chaos?* Bathurst (NSW): Crawford House Press.

Laksono, P.M., 1990. Wuut Ainmehe Nifun, Manut Ainmehe Tilor [Eggs from One Fish and One Bird]: A Study of the Maintenance of Social Boundaries in the Kei Islands. Ithaca: Cornell University (Ph.D. thesis).

————, 2004. 'Benih-Benih Perdamaian Dari Kepulauan Kei [The Seeds of Peace from Kei Archipelago].' In P.M. Laksono and R. Topatimasang (eds), op. cit.

Laksono, P.M and R. Topatimasang (eds), 2004. *Ken sa Faak: Benih-Benih Perdamaian dari Kepulauan Kei[Ken sa Faak: Seeds of Peace from the Kei Islands]*. Jogyakarta: Insist Press.

Lasomer, X., 1985. 'De Kei-Eilanden: De Uitbreiding van de Nederlanse Invloeding in de Tijd van de Ethisch Politiek [The Kei Islands: The Expansion of Dutch Influence in the Time of Ethical Politics].' Nijmegen: Catholic University (unpublished manuscript).

Lawalata, J., 1969. No Title. Unpublished paper about the author's experience in the Kei Islands from 1914 to 1939. Ambon: Rumphius Library.

Lokollo, J.E., 1988. 'Hukum Sasi di Maluku: Suatu Potret Binamulia Lingkungan Pedesaan yang Dicari Pemerintah [The Sasi Law in Maluku: A Portrait of Rural Environmental Management as Requested by the Government].' Ambon.

————, 1994. 'Asas-Asas Hukum Adat Kelautan dan Manfaatnya Bagi Pembinaan Peraturan Daerah di Kabupaten Maluku Tengah Dalam Rangka Implementasi Undang-Undang Nomor 4 tahun 1982 dan Undang-Undang Nomor 9 tahun 1985 [The Foundation of Marine Traditional Law and Its

Function in Supporting Regional Regulation for the Implementation of Laws No. 4 of 1982 and No. 9 of 1985 in Central Maluku].' Ambon: University of Pattimura, Faculty of Law.

Marsono, 1980. *Undang-Undang Republik Indonesia Nomor 5 tahun 1979 tentang Pemerintahan Desa [The Indonesian Republic Law No. 5 of 1979 on Village Government].* Jakarta: Ichtiar Baru.

Marut, D.K., 2004. 'Petuanan dan Sasi: Hak Komunal dan Manajemen Sumberdaya Alam di Maluku [*Petuanan* and *Sasi*: Communal Rights and Natural Resource Management in Maluku].' In P.M. Laksono and R. Topatimasang (eds), op. cit.

Matsuda, Y. and Y. Kaneda, 1984. 'The Seven Greatest Fisheries Incidents in Japan.' In K. Ruddle and T. Akimichi (eds), op. cit.

McCay, B.J., 1995. 'Common and Private Concerns.' *Advances in Human Ecology* 4: 90–116.

McCay, B.J. and S. Jentoft, 1996. 'From Bottom Up: Participatory Issues in Fisheries Management.' *Society and Natural Resources* 9: 237–250.

———, 1998. 'Market or Community Failure? Critical Perspectives on Common Property Research.' *Human Organization* 57: 21–29.

McCay, B.J. and J.M. Acheson (eds), 1987. *The Question of the Commons: The Culture and Ecology of Communal Resources.* Tucson: University of Arizona Press.

McGoodwin, J.R., 1990. *Crisis in the World's Fisheries: People, Problems, and Policies.* Stanford: Stanford University Press.

Milan, V., 1993. 'Cyanide Fishing, Tubbataha Reefs and the Chinese Connection.' *Coastal Management in Tropical Asia* 1: 16–19.

Monk, K.A., Y. de Fretes and G. Reksodiharjo-Lilley (eds), 1997. *The Ecology of Nusa Tenggara and Maluku.* Singapore: Periplus Editions (Ecology of Indonesia Volume 5).

Ngamelubun, M., 2004. 'Perempuan dalam Resolusi Konflik dan Rekonsiliasi di Kei [Women in Conflict Resolution and Reconciliation in Kei].' In P.M. Laksono and R. Topatimasang (eds), op. cit.

Nielsen, J.R. and T. Vedsmand, 1997. Fishermen's organisations in fisheries management - Perspectives for fisheries co-management based on Danish fisheries. *Marine Policy* 21, No. 2: pp. 277-288

Nikijuluw, V.P.H., 1994. 'Indigenous Fisheries Resource Management in the Maluku Islands.' *Indigenous Knowledge and Development Monitor* 2(2): 6–8.

Novaczek, I., I.H.T. Harkes, J. Sopacua and M.D.D. Tatuhey, 2001. 'An Institutional Analysis of *Sasi Laut* in Maluku, Indonesia.' Penang: International Center for Living Aquatic Resources Management (Technical Report 59).

Novaczek, I., J. Sopacua and I. Herkes, 2001. 'Fisheries Management in Central Maluku, Indonesia 1997–98.' *Marine Policy* 25: 239–249.

Ohoitimur, Y., 1983. Beberapa Sikap Hidup Orang Kei: Antara Ketahanan Diri dan Proses Perubahan [Some Attitudes of the Kei Population: Between Preservation and Change]. Pineleng (Manado): Sekolah Tinggi Seminari [Advanced Seminary].

Ostrom, E., 1990. *Governing the Commons: The Evolution of Institutions for Collective Action.* Cambridge: Cambridge University Press.

Pannell, S., 1993. 'Circulating Commodities: Reflections on the Movement and Meaning of Shells and Stories in North Australia and Eastern Indonesia.' *Oceania* 64: 57–76.

———, 1997. 'Managing the Discourse of Resource Management: the Case of *Sasi* from "Southeast" Maluku, Indonesia.' *Oceania* 67: 289–307.

Persoon, G.A. and D. M. E. Van Est, 2003. 'Co-Management of Natural Resources: The Concept and Aspects of Implementations.' In G.A. Persoon, D.M.E. van Est and P.E. Sajise (eds), *Co-Management of Natural Resources in Asia: A Comparative Perspective.* London: Taylor & Francis.

Peterson, N. and B. Rigsby, 1998. 'Introduction.' In N. Peterson and B. Rigsby (eds), *Customary Marine Tenure in Australia.* Sydney: University of Sydney.

Pinkerton, E., 1989. 'Introduction: Attaining Better Fisheries Management through Co-Management: Prospects, Problems, and Propositions.' In E. Pinkerton (ed.), *Co-Operative Management of Local Fisheries: New Directions for Improved Management and Community Development.* Vancouver: University of British Columbia Press.

Pollnac, R.B., 1984. 'Investigating Territorial Use Rights among Fishermen.' In K. Ruddle and T. Akimichi (eds), op.cit.

Polunin, N.V.C., 1984. 'Do Traditional Marine "Reserves" Conserve? A View of Indonesian and New Guinean Evidence.' In K. Ruddle and T. Akimichi (eds), op.cit.

Pomeroy, R.S., (ed.), 1994. *Community Management and Common Property of Coastal Fisheries in Asia and the Pacific: Concepts, Methods and Experiences*. Manila: International Center for Living Aquatic Resources.

Rahail, J.P., 1993. 'Larvul Ngabal: Hukum Adat Kei, Bertahan Menghadapi Arus Perubahan [Larvul Ngabal: The Kei Customary Law, Resistance to Change].' Jakarta: Sejati Foundation.

———, 1995. Bat Batang Fitroa Fitnangan: Tata Guna Tanah dan Laut Tradisional Kei. [Bat Batang Fitroa Fitnangan: Traditional Land and Marine Use Patterns in Kei.] Jakarta: Sejati Foundation (Seri Pustaka Khasanah Budaya Lokal Volume 4).

Rahawarin, A., 1959. Sejarah Nama-Nama Pemegang Kekuasaan Pemerintahan dalam Daerah Ubohoifak/Englarang Sejak Zaman Purbakala s/d Kedudukan Penjawat di Dullah/Pulau-Pulau Kei (Componi) Sampai Kini di Englarang [The History of the Government Power Holder's Names in the Territory of Ubohoifak/Englarang from Former Times to the Establishment of the Dutch (Company) Post-Holder in Dullah/Kei Islands and Continuing Now in Englarang.] Kei Besar.

Reid, A., 1983. '"Closed" and "Open" Slave Systems in Precolonial Southeast Asia.' In A. Reid (ed.), *Slavery, Bondage and Dependency in Southeast Asia*. St Lucia: University of Queensland Press.

Renyaan, P.H., 1990. *Sejarah Adat Kei [The History of Kei Tradition]*. Langgur.

———, 1996. *Seratus Tahun Perkembangan Agama Katolik di Kepulauan Kei, 1889–1989[A Century of Catholic Development in Kei Island, 1889–1989]*. Ambon: Pusat Pengembangan Pastoral Keuskupan Amboina [Amboin Diocese Pastoral Development Center].

Renwarin, F., n.d. 'Catatan Ringkas Sejarah Raja-Raja di Kepulauan Kei [A Brief History of Kings in Kei Archipelago].' Tual.

Riedel, J.G.F., 1886. *De Sluik-en Kroesharige Rassen tusschen Selebes en Papua[Illicit and Frizzy Haired Breeds between Celebes and Papua]*. Gravenhage: M. Nijhoff.

Rubec, P.J., 1986. 'The Effects of Sodium Cyanide on Coral Reefs and Marine Fish in the Philippines.' In J.L. Maclean, L.B. Dizon and L.V. Hosillos (eds), *The First Asian Fisheries Forum*. Manila: Asian Fisheries Society.

———, 1988. 'Cyanide Fishing and the International Marine Life Alliance Net-Training Program.' *Tropical Coastal Area Management* 3: 11–13.

Ruddle, K. and T. Akimichi (eds), 1984. *Maritime Institutions in the Western Pacific*. Osaka: National Museum of Ethnology.

Ruddle, K. and A. Satria (eds), 2010. *Managing Coastal and Inland Waters: Pre-Existing Aquatic Management Systems in Southeast Asia*. Dordrecht: Springer.

Samego, I., I. N. Bhakti, H. Sulistyo, R. Sihbudi, M. H. Basyar, M. Nurhasim, N. I. Subono and S. Yanuarti, 1998. *Bila ABRI Berbisnis: Buku Pertama yang Menyingkap Data dan Kasus Penyimpangan dalam Praktik Bisnis Kalangan Militer [When ABRI Do Business: The First Book Uncovering Data and Illegal Cases in Military Business Practices]*. Bandung: Mizan.

Satria, A. and D.S. Adhuri, 2010. 'Pre-Existing Fisheries Management Systems in Indonesia, Focusing on Lombok and Maluku.' In K. Ruddle and A. Satria (eds), op. cit.

Satria A. and Y. Matsuda, 2004. 'Decentralization Policy: An Opportunity for Strengthening Fisheries Management System?' *Journal of Environment & Development* 13: 179–196.

Schreurs, P.G.H., 1992. *Lanjutan Karya St Fransiskus Xaverius: Kebangkitan Kembali Misi Katolik di Maluku 1886–1960 [The Continuation of the Work of St Fransiskus Xaverius: The Resurrection of the Catholic Mission in Maluku 1886–1960]*. Ambon: Pusat Pengembangan Pastoral Keuskupan Amboina [Amboin Diocese Pastoral Development Center].

Scott, A., 1955. 'The Fishery: The Objective of Sole Ownership.' *Journal of Political Economy* 63: 116–124.

Silubun, E., 2004. '"Ken sa Faak": Kerangka Kerja Rekonsiliasi dan Pengungkapan Kebenaran Menurut Adat Kei ["Ken sa Faak": The Framework for Reconciliation and Uncovering the Truth According to Kei Tradition].' In P.M. Laksono and R. Topatimasang (eds), op. cit.

Soselisa, H.L., 2002. Memories and Fragments: Resource Management in Central Maluku, Eastern Indonesia. Darwin: Northern Territory University (Ph.D. thesis).

Thorburn, C.C., 2000. 'Changing Customary Marine Resource Management Practice and Institutions: The Case of *Sasi Lola* in the Kei Islands, Indonesia.' *World Development* 28: 1461–1479.

————, 2001. 'The House that Poison Built: Customary Marine Tenure Property Rights and the Live Food Fish Trade in the Kei Islands, Southeast Maluku. *Development and Change* 32: 151–180.

————, 2005. '*Musibah*: Governance, Intercommunal Violence and Reinventing Tradition in the Kei Islands, Southeast Maluku.' Clayton (VA): Monash University, Centre of Southeast Asian Studies (Working Paper 125).

Topatimasang, R., 2004. 'Pengantar: Toil u Ne It Savhak Muir [Introduction: To Consider the Consequences of Every Action].' In P.M. Laksono and R. Topatimasang (eds), op. cit.

UPPPSL (Universitas Pattimura, Pusdi-PSL), 1995 . 'Kajian Hukum Tentang Norma Adat dalam Perlindungan Lingkungan [A Study of Customary Norms about Environmental Protection].' Ambon: UPPPSL [Pattimura University, Environmental Studies Centre].

Valeri, V., 1989. 'Reciprocal Centres: The Siwa-Liwa System in the Central Moluccas.' In D. Maybury-Lewis and U. Almagor (eds), *The Attraction of Opposites: Thought and Society in the Dualistic Mode*. Ann Arbor: University of Michigan Press.

Van Hoëvell, G.W.W.C., 1890. 'De Kei-Eilanden [The Kei Islands].' *Tijdschrift voor het Indische Taal-, Land- en Volkenkunde* 33: 102–159.

Van Wouden, F.A.E., 1968. *Types of Social Structure in Eastern Indonesia* (transl. R. Needham). The Hague: Martinus Nijhoff.

Van Klinken, G., 2001. 'The Maluku Wars: Bringing Society Back In.' *Indonesia* 71: 1–26.

Vatikonis, M.R.J., 1998. *Indonesian Politics under Suharto: the Rise and Fall of the New Order* (3rd edition). London: Routledge.

Von Benda-Beckmann, F., K. von Benda-Beckmann and A. Brouwer, 1992. 'Changing "Indigenous Environmental Law" in the Central Moluccas: Communal Regulation and Privatization of *Sasi*.' Paper read at the Congress of the Commission on Folk Law and Legal Pluralism, Victoria University, Wellington, 21–24 August.

Wahyono, A. A.R. Patji, D.S. Laksono, R. Indrawasih, Sudiyono, and S. Ali, 2000. *Hak Ulayat Laut di Kawasan Timur Indonesia/Communal Marine Tenure in Eastern Indonesia]*. Yogyakarta: Media Pressind.

Wallace, A.R., 1986. *The Malay Archipelago: The Land of the Orang-Utan, and the Bird of Paradise*. Oxford: Oxford University Press.

Warren, C. and K. Elston, 1994. 'Environmental Regulation in Indonesia.' Perth: Murdoch University, Asia Research Centre (Asia Paper 3).

Zerner, C., 1991. 'Imagining the Common Law in Maluku: Of Men, Molluscs, and the Marine Environment.' Paper presented at the second annual meeting of the International Association for the Study of Common Property, Manitoba, 26–29 September.

———, 1992. 'Community Management of Marine Resources in the Maluku Islands.' Paper presented at the FAO/Japan expert consultation on the 'Development of Community-Based Coastal Fishery Management Systems for Asia and the Pacific', Kobe, Japan, 8–12 June.

———, 1994a. 'Through a Green Lens: The Construction of Customary Environmental Law and Community in Indonesia's Maluku Islands.' *Law & Society Review* 28: 1079–122.

———, 1994b. 'Transforming Customary Law and Coastal Management Practices in the Maluku Islands, Indonesia (1870–1992).' In D. Western and R.M. Wright (eds), *Natural Connections: Perspectives in Community-Based Conservation*. Washington (DC): Island Press.

———, 1996. *Sea Change: The Role of Culture, Community, and Property Rights in Managing Indonesia's Marine Fisheries*. Jakarta: Obor Foundation.